COSMOPOLITAN SCIENTISTS

CULTURE AND ECONOMIC LIFE

COSMOPOLITAN SCIENTISTS

How a Global Policy of Commercialization Became Japanese

NAHOKO KAMEO

STANFORD UNIVERSITY PRESS
Stanford, California

Stanford University Press
Stanford, California

Printed in the United States of America on acid-free, archival-quality paper

Library of Congress Cataloging-in-Publication Data
Names: Kameo, Nahoko, author.
Title: Cosmopolitan scientists : how a global policy of commercialization became Japanese / Nahoko Kameo.
Other titles: Culture and economic life.
Description: Stanford, California : Stanford University Press, 2024. | Series: Culture and economic life | Includes bibliographical references and index.
Identifiers: LCCN 2024013692 (print) | LCCN 2024013693 (ebook) | ISBN 9781503639928 (cloth) | ISBN 9781503640405 (paperback) | ISBN 9781503640412 (ebook)
Subjects: LCSH: Research—Economic aspects—Japan. | Universities and colleges—Research—Economic aspects—Japan. | Academic-industrial collaboration—Japan. | Globalization—Economic aspects—Japan.
Classification: LCC Q180.55.E25 K36 2024 (print) | LCC Q180.55.E25 (ebook) | DDC 001.40952—dc23/eng/20240412
LC record available at https://lccn.loc.gov/2024013692
LC ebook record available at https://lccn.loc.gov/2024013693

Cover design: Hollis Duncan

Contents

Acknowledgments

Writing a book takes a long, long time. Of course, it is probably a good thing that I didn't know this, because if one knew that it would take about double the time that they planned for, fewer books would be written. Developing a book was a long journey, and so many people helped me personally and intellectually. First, thanks to institutions (as this is a book on new institutional theory). I thank the University of California–Los Angeles (UCLA) Department of Sociology and New York University Department of Sociology for all their functions, such as the 237 seminar (a legendary seminar at UCLA) and the junior faculty working group at New York University (nice and smart assistant professors hanging out and discussing one another's work). They gave me robust and resilient scaffoldings to support and build my own academic work. The UCLA Terasaki Center for Japanese Studies, National Science Foundation (Award Number 1063988), and Ewing Marion Kauffman Foundation have supported this project, and I thank them for taking a risk on me. Second, so many friends and colleagues took the time to sit down with me and offered me their invaluable thoughts and comments on this project. I would like to thank Maria Abascal, Gianpaolo Baiocchi, Alex Barnard, Beth Bechky, Bart Bonikowski, Kate Choi, Paul DiMaggio, Nina Eliasoph, Anne-Laure Fayard, Jacob Foster, Tim Hallett, Jack Katz, Carly Knight, Pamela Prickett, Gabriel Rossman, Anna K. Skarpelis, Iddo Tavory, Stefan Timmermans, Jack Whalen, Marilyn Whalen, and Lynne G. Zucker. Special thanks to Lynne G. Zucker, who supported me throughout the years of graduate training and who was relentlessly enthusiastic and optimistic about my academic journey. Third, I thank all the scientists who agreed to be interviewed for this study. Their openness, intellect, and sense of responsibility truly inspired me. I tried my best to hear what they had to say and represent it in a sociological manner. I hope I succeeded.

When Marcela Maxfield, the editor of this book, met me, the book was still in its early stages. Marcela has been so encouraging and patient on the intellectual side, and so consistent and dependable on the process side. I feel incredibly lucky to have such an amazing editor. Kelly Besecke helped me clarify my writing and correct every single sentence I wrote that was vague, nongrammatical, or overly complicated. I learned so much about writing, and writing a book, from them. I also thank the anonymous reviewers who gave very constructive criticisms on the draft. Earlier versions of chapter 4 and chapter 6 have been previously published in *Theory and Society* (see Kameo 2015; 2024.)

This book would not have been written without my family's continuous support. I thank my two daughters, Eliana and Amalya, because they are the very best. Iddo read so many of my drafts he could probably recite the book. My parents-in-law—Eva Jablonka, Doron Tavory, and Ada Ushpiz—also took a serious interest in this book, not because it was their daughter-in-law writing it, but because they are intellectually curious and ambitious. I thank them for not only asking me when the book will be published but asking me about the intellectual contributions that I am making. My parents, Ruriko and Kyoji Kameo, gave me full support with the kind of patience only parents can afford. I am one of those kids who are so extremely lucky that they do not think about their parents too much, precisely because I have always been confident about their love and acceptance. This book is dedicated to them.

ONE Global Policy, Inhabited Institutionalism, and Commercialization of Research

ON OCTOBER 1, 2018, the Nobel Foundation announced that year's recipients of the Nobel Prize in Physiology or Medicine. James P. Allison and Tasuku Honjo were awarded the prize "for their discovery of cancer therapy by inhibition of negative immune regulation." Later that day, a press conference was held at Kyoto University, where Dr. Honjo worked, to celebrate this accomplishment. Dr. Honjo opened his remarks,

> I am extremely happy and honored to receive the Nobel Prize in Physiology or Medicine. I am earnestly indebted to my hardworking collaborators, students, people who supported me in various ways, and my family, who supported me for such a long time. I cannot describe how grateful I am to so many people. My extremely basic research led to the discovery of PD-1 in 1992, and the following research created the clinical application, new cancer immunotherapy. Occasionally, people come to me to say this treatment saved them from critical illness—they have recovered, thanks to me. That is when I palpably feel that my research is truly meaningful; nothing makes me happier. I am such a lucky person to receive the prize on top of this. I plan to continue research for a while myself, together with many researchers around the world, in order for the immunotherapy to save even more cancer patients. With our efforts, I expect the therapy to become more effective in the future. This Nobel Prize was given to my research that was fundamental,

> that then developed into clinical applications. If the fact that the prize was awarded to such research could accelerate the development of basic medical research and encourage many basic researchers, it would be more than I could ever dream of.[1]

The Nobel Prize was originally established with money earned from the invention of dynamite—a history that speaks to a key dilemma of scientific inquiry. On one hand, scientists are supposed to produce knowledge for its own sake—pure scientific inquiry for society. On the other hand, the most prestigious scientific award in the world only exists because of money made by a practical product generated by applied science: dynamite. In this tension between basic science and its commercialization, Dr. Honjo's research is a scientist's dream: it is outstanding scholarship that reveals fundamental mechanisms about the human body, and it is also *useful*—in fact, people come to him to tell him that his research saved their lives. PD-1 is a protein on the surface of cells that plays a role in regulating the immune system. It took more than two decades after the discovery of PD-1 for immunotherapy based on Dr. Honjo's research to come to market in Japan and around the world. His opening remarks at the press conference placed great emphasis on the importance of basic research—and funding for it. The speech concludes by calling for more basic research and a research environment that supports it. Especially since his research had yielded a lucrative biomedical application, Dr. Honjo needed to emphasize that this application had only come about as a result of decades-long basic research about cell behavior.

The management of the boundary and the possible synergy between academic research and commercialization has become an important question for policymakers and a practical problem for university scientists. Over the past decade, only 55% of public research funding for universities in Japan has been targeted to basic research, and more and more government funding sources require researchers to articulate the potential for marketable applications of their research. Dr. Honjo's remarks were based on a sense of urgency in an effort to counter this pressure for application. In fact, after the initial discovery of PD-1, it took many years, publications, patents, and labs in addition to Dr. Honjo's to finally produce the medical application. He applied for a patent on PD-1 in 1995,[2] and it took nearly twenty years and many additional patents for Nivolumab (trade name: Opdivo) to be marketed by Ono Pharmaceutical, a midsized Japanese pharmaceutical company.[3] By 2018, more than sixty countries had approved Nivolumab for the treatment of various cancers, including metastatic melanoma and non–small cell lung cancer.

As Dr. Honjo continued on to the question-and-answer section of the press conference, he kept emphasizing the importance of focusing on the fundamental mechanisms of the human body while keeping an eye on the possibility of medical application. When a journalist asked him about "the partnership between [him] and Ono Pharmaceutical," he expressed reservations about his relationship with the firm:

> I am a scientist; I do not have a business. Partnership would refer to a relationship between firms. My relationship with Ono is that I gave Ono the license in terms of the intellectual property. [. . .] Regarding this research, Ono Pharmaceutical did not contribute to the research itself. That is very clear. And Ono is being licensed the patent, so I would like them to give back to the university sufficiently. By that, [I mean that] rather than me doing new research, my hope is that there will be a foundation based on reciprocation to Kyoto University and that the next generation of scholars will be encouraged, nurtured by that, creating new seeds. And that then cycles back to Japanese pharmaceuticals. I hope to create this kind of win-win relationship, and that's what I have been asking Ono to do for a long time.[4]

In these comments, Dr. Honjo reveals surprising bitterness about his collaboration with industry and an attachment to a particular form of reciprocation. He makes it clear that he wants to create a foundation so that research "cycles back to Japanese pharmaceuticals" such as Ono, but he also expresses reservations about his relationship with the firm. He explicitly mentions that Ono did not contribute to the research that led to the biomedicine and suggests that Ono should contribute more to Kyoto University.

This, as it turns out, was the beginning of a lengthy series of legal battles between Dr. Honjo and Ono Pharmaceutical. At the core of this legal saga was Dr. Honjo's conviction that when the drug became so successful, Ono should have voluntarily contributed to Kyoto University and that his original agreement with Ono offered too little compensation in light of how lucrative the drug became for the firm.[5]

Yet in fact, the original agreement between Dr. Honjo and Ono had explicitly precluded large monetary compensations to Dr. Honjo and made no mention of any financial obligation at all to Kyoto University. With such an agreement in place, why did Dr. Honjo and Ono end up in such protracted legal battles over Ono's financial responsibilities? More perplexingly, why did Dr. Honjo win these battles? Where did this notion that a pharmaceutical company should "give back" to a university come from? And why was that notion not relevant when Dr. Honjo and Ono Pharmaceutical started collaborating in the 1980s?

The answer to this puzzle lies in the unique history of Japanese universities' relationships with Japanese firms. Before the 1990s, Japanese universities collaborated with Japanese firms according to a particular routine of reciprocity. Because universities were thought to contribute to public knowledge, their interactions with industry were largely kept informal. In the late 1990s, however, this tradition began to change under the influence of the United States. In the 1980s, U.S. universities had changed the policies that governed faculty inventions and entrepreneurship to encourage explicit, formal, and aggressive commercialization of academic knowledge—especially in biotechnology. In the 1990s, Japan started to adopt similar policies and reshape how Japanese university scientists interacted with industry.

Dr. Honjo's case illuminates both continuities and changes amid institutional transition. His patents for PD-1 are all joint patents with Ono Pharmaceutical, listing both Dr. Honjo and Ono Pharmaceutical as joint assignees. As this book will show, this was one of several possible arrangements that were quite common before the science and technology policy change that occurred between 1995 and 2005. Dr. Honjo's remarks about Ono during his press conference implied that Ono basically free-rode on the patent. But Ono is a co-assignee of the patents, so, legally speaking, it has full rights to exercise the patents—after all, it co-owns them. Dr. Honjo and Ono Pharmaceutical have undisclosed contracts that establish compensation to Dr. Honjo as an individual, and as long as those terms are fulfilled, Dr. Honjo doesn't have a legal right to ask Ono to contribute to Kyoto University. In the press conference, then, Dr. Honjo was sharing his personal sense that his arrangement with Ono—one he'd chosen to engage in—was unfair.

Dr. Honjo wanted to leverage his Nobel Prize to create a research fund at Kyoto University. His frustration with Ono came from the fact that the company had earned astounding levels of sales revenue—about $803 million in 2017 in Japan, licensing revenue overseas excluded—from this single medicine alone but had not yet indicated how much it would donate to the university's research fund.[6] Dr. Honjo's position makes sense from today's perspective, now that Japan has fully transitioned to the American model, whose formal rules assign universities the ownership of their faculty members' inventions. It also, perhaps, made sense personally, as Dr. Honjo came to serve as the chairperson of Japan's Council for Science and Technology, the body that directly advises the prime minister on science and technology policy. His expectations were grounded on the new logic that took root in Japanese academia as a result of new policies. According to this logic, universities own the patents to their faculty members' inventions, and when those inventions result in marketable products, the university is to be rewarded financially for the active, formal role it has played in enhancing innovation.

Ono Pharmaceutical seems to have been surprised by Dr. Honjo's resentment. The president of Ono said in an interview that Dr. Honjo's criticism had been "unexpected," that the company had paid Dr. Honjo everything that was due to him, and that they "have been helping Dr. Honjo's research since the 1980s." The firm, he emphasized, had continued to support Dr. Honjo despite internal pushback. Cancer immunotherapy had initially been thought too risky to gamble on, and the firm had only continued research on PD-1 because "one employee was enthusiastic about Dr. Honjo's research." The president continued, "Opdivo's R&D number is 4538. In our firm, the correct number for an R&D project for a cancer drug must start with 7. He started the project with a project number that started with 4, which meant 'R&D for other drug candidates,' so the management did not realize [that the research was for a cancer drug]. It helped that we have a culture to tackle a challenge that nobody else is willing to."[7] The president's account emphasized the long-standing collaboration between Ono and Dr. Honjo, the firm's investment in his research at a time when it had been seen as a risky investment, and the central role of an employee who was so personally enthusiastic about the research that he hid its real subject matter so that Ono would continue to fund it. Setting aside the contradictory claims that Ono's culture welcomed challenges but would have stopped funding the project if its actual subject matter had been known, the president's account highlights passion, trust, and the company's personal relationship with Dr. Honjo; Kyoto University isn't even part of the story.

Again, the differences in commercialization practices over time allow us to understand the position of Ono Pharmaceutical's president. His perspective was not unusual. Until the 1980s, Japan had no formal, systematic way for firms to collaborate with university professors. As a result, Japanese academia often interacted with firms based on an informal system called *shogaku-kifu*, which translates to "scholarship donations"—firms would donate money for research to a specific lab in a Japanese university and collaborate on a research project. If and when there were some inventions, the professors would let the firm take intellectual property rights on them—with the informal understanding that the firm would continue to fund the lab. This informal method of collaboration remained common into the 1990s and early 2000s. Until the late 1990s, it was very rare for universities to own the patents for faculty inventions. Dr. Honjo therefore had to rely on Ono to file and maintain intellectual property rights to his own research products. But from his perspective, Ono's intellectual contribution had been minimal: he had simply "licensed" the patent to Ono (in reality, Ono co-owns the patents, so the use of the word "license" is not precise). In Honjo's eyes, Ono's collaboration with him at Kyoto University and their minimal intellectual contribution meant that, ethically, the firm owed the university some form of return.

For anyone accustomed to the current American system of university-firm collaboration and commercialization, it is perhaps easy to take Dr. Honjo's side. Japanese firms may seem to have been exploiting university scientists, and Japan's transition to the American-style system may seem an obvious move. And in fact, some critics have argued that Japanese firms were used to treating university scientists as free-of-charge sources of knowledge and innovation. But to understand the changes in the Japanese system and the repercussions of these changes, it is critical to see the conflict from a historical perspective. Until the 1990s, Japan's universities weren't supposed to be generating patentable innovations. Universities were regarded as places to "research and educate," and activities such as innovation and patenting were considered outside the university's official scope.[8] As a result of universities' hands-off approach to commercialization, it was fairly common for professors to give firms the patents to their inventions as late as the early 2000s. Many, if not all, professors took this arrangement for granted.

Underlying this kind of relationship with the firm was a different approach to university-firm collaborations—one that had been common in Japan before the policy changes of the late 1990s. This approach was characterized by gift-exchange-like practices and valued long-term relationships that obscured the issue of monetary compensation. When Ono's president said, "We have been helping Dr. Honjo's research since the 1980s," he was evoking precisely this kind of relationship. Now that the new American-style policies have been introduced, is this old style of collaboration lost, and is Dr. Honjo the harbinger of the victory of the new approach? Or do aspects of the old system endure despite the policy changes—and if so, why? What will be the new picture of university-firm collaboration in Japanese academia, and how do scientists play a part in shaping it?

When a global policy spreads and is adopted, only one country gets to be the first to create it. The United States' new policies to regulate university-firm interactions and commercialization of research were developed at the same time that such practices were becoming common. In Japan, however, university-firm interactions had existed for quite a while before any policies were developed to regulate them. As a result, university scientists and firms developed their own informal system based on building long-term relationships of trust. When U.S.-style policies were imported in the late 1990s and 2000s, they were largely exogenous and different from what scientists considered to be appropriate ways to govern commercialization. The new policies asked scientists and firms to move from informal trust relationships to a system of formal procedures and contracts. According to the new policies, patents must be filed and owned by the university, and university administrators must be involved. Research collaborations, faculty entrepreneurship, and technology transfer are incessantly encouraged and recorded in government statistics and must be formally compensated.

A good example of the older, trust-based style of collaboration is the story of Dr. Koyama, a fiber engineering professor and the president of a national research university in Japan. In 2017, Dr. Koyama described his experience of collaboration with a firm in an article he wrote for the *Sangakukan renkei Janaru* (Industry-University-Government Collaboration Journal), the official journal of the Japan Science and Technology Agency, a government subsidiary that manages funding and administration for many R&D programs. According to his article, a senior director at an equipment manufacturing company visited Dr. Koyama's lab one day, and the two men began to develop an informal collaboration. The relationship between the professor and the company continued on and off for years without any formal or even verbal agreement about the collaboration. Almost a decade after the initial contact, the firm created new equipment based on the collaboration. Later that year, the professor was invited to the firm's headquarters in Tokyo, where Dr. Koyama's collaborator, who had by then become the firm's president, said he would like to show his gratitude to Dr. Koyama and asked, "Can you please tell me your bank account?" Dr. Koyama firmly refused, mentioning that he would lose his job if he received money like that. The president insisted, and they agreed that the firm could set up research collaborations aimed at continuing to improve the equipment. Every year from then on, the firm sent Dr. Koyama research funds that became an essential part of his research budget.

In his article, Dr. Koyama recalled this experience as "a collaboration that was solely based on 'trust' and can be called 'Japanese culture.'"[9] He expressed astonishment, looking back, that the firm compensated his lab without even a verbal promise, let alone a contract. And in Dr. Koyama's story, once again, there is absolutely no mention of his university—or of payment—until the research fund for new equipment was created.

These days, Dr. Koyama is the president of a research university with the responsibility to propel research collaborations in a more formal manner. But in his article, he seems to speak of the old approach to collaboration with approval and as a memory of good times. He attributes this approach to Japanese culture: it was essentially Japanese, he says, to build a collaboration based on trust. This reference to Japan was not unique: in my interviews, Japanese university scientists evoked national sentiments repeatedly and consistently, not only when they spoke of the old system, but even more so when they described how Japan had adopted and altered U.S.-originated commercialization policies. Even Dr. Honjo, whose frustration with Ono Pharmaceutical derived from having worked with the firm under the old Japanese system, framed his criticism of Ono as emerging from his wish for university research to "cycle back to *Japanese* pharmaceuticals." Essentially, the stories of Dr. Honjo and Dr. Koyama are the story of a larger transition: the transition between the older, Japanese method of managing relationships between academia and industry and the new American-style system of research commercialization. And Dr. Honjo and Dr. Koyama are far from alone: many university scientists in Japan lived through this period of institutional and cultural change.

This book is about how a global policy—the U.S.-originated commercialization policy—was adopted in Japan and how we can understand continuity and change in the ways scientists do science and commercialization. Statistical findings presented in the next chapter clearly show that the policy change resulted in changes in the intended direction. The number of university-firm collaborations, university patents, and faculty-led startups increased after the policy change. However, these statistics fail to capture the whole picture. This larger picture includes other quantitative changes, but it also includes qualitative changes in how Japanese university scientists collaborated with industry and commercialized their inventions and how they felt about these changes in policy and practice.[10]

Existing literature on this topic looks only at the macro-level trends. Unsatisfied with this literature, I visited Japanese university bioscientists in their labs to find out how they worked with industry, dealt with patents, and got involved in entrepreneurship. I asked them how they had come to interact with industry and involve themselves with research commercialization and, for the many who had collaborated with firms both before and after the policy change, how they had changed their practices throughout their careers. I also conducted archival research and reviewed patent statistics to understand how Japanese university bioscientists had responded to the new imported structure of commercialization (for methodological details, see the appendix). This additional research provided valuable context that supplemented this book's primary focus, which is Japanese university scientists' perceptions, actions, and meaning-making. My analysis shows how these scientists adapted the new policy to their established ways of doing things, how they simultaneously imported practices that they had learned from their international experiences, and how they creatively used national identity to localize global policy. Showing the processes through which global policy is adopted, the book shows that the diffusion of global policy is a local accomplishment that depends on active cultural work done by the local actors whose lives are most affected by the policies.

Inhabited Institutionalism as a "New" New Institutional Theory

The university has undergone significant changes in its substance during the twenty-first century. Historically developed as a religious and then an elite institution for education and research, the university in the United States (and many advanced economies) had become a site for mass education by the mid-1970s. But in the 1980s, just as the university had established itself as an icon of middle-class upward mobility, another change in its prescribed purpose began to materialize. In addition to the dual missions of education and research, policymakers were starting to recognize a third mission: fostering commercial innovation. The present-day university is expected to directly contribute to the commercialization of the knowledge it produces. Almost every essential invention, we are told, has its origin in the university: weather-resistant crops, safety equipment, computers, the internet, biomedicine, robots—the list goes on and on. Every reputable U.S. research university is equipped with a specialized technology transfer office that celebrates its ability to parlay academic knowledge into real-world innovations. In 2016 alone, 7,021 U.S. patents were issued to U.S. universities, and total licensing income was about three billion U.S. dollars.[11]

The commercially "innovative" university is a recent invention, and it has a distinctly American flavor. As I show in the next chapter, commercialization-oriented academia is a result of three forces simultaneously at work in the 1970s and early 1980s: the birth of biotechnology, changes in intellectual property rights, and other policy and organizational changes based on the neoliberal idea that a university's value depends on its economic impact. The reform of intellectual property rights gained momentum with the passage of the Bayh-Dole Act in 1980, which allowed universities to own faculty inventions. The Bayh-Dole Act radically streamlined the process for granting a university ownership of intellectual property arising from professors' research. The number of faculty inventions, the number of university-related startups, and patent licensing revenue all rose rapidly at most research universities in the U.S.[12] Meanwhile, universities established technology transfer offices and began to maintain a staff of professional intellectual property (hereafter IP) managers. Together, these legal and institutional changes solidified the emergence of a more commercialization-oriented and entrepreneurial form of academia.

The success of university commercialization of research in the U.S. drew the attention of other countries, and by the end of the twentieth century, many countries with strong science and technology sectors were looking to imitate what the U.S. had done to promote the commercialization of university research. Countries around the world adopted the United States' legal and institutional structure for the commercialization of research, including the promotion of faculty inventions and entrepreneurship, legal and institutional changes that enable university ownership of intellectual property, and streamlined formal pathways for firms to collaborate with university scientists.

This is the kind of phenomenon that new institutional theory set out to explain. New institutional theory is a sociological approach to understanding organizational behavior in the context of a changing environment. Why, new institutionalist scholars ask, do institutions develop, and how do particular practices and arrangements become "just the way it is" in the first place? Why do different organizations develop similar forms and routines despite their differences? Conventional organizational theory might attribute institutional imitation to the form's efficiency (for example, the creation of a diversity management office might produce a better workforce) or to power relations (for example, if a powerful country promotes a particular standard, others are compelled to follow). By contrast, new institutional scholars look at *cultural* forces that drive the adoption of a form.

In one of their canonical papers, new institutionalists John W. Meyer and Brian Rowan argue that an organization adopts "institutionalized products, services, techniques, policies, and programs" not because they will make the organization more efficient but because doing so increases the organization's *legitimacy*.[13] One good example is the education sector. Schools tend to adopt surprisingly similar measures despite considerable variation in their internal structures, the populations they serve, and the ambiguous effectiveness of the measure in question.[14] Other new institutional scholars have developed this line of thought. Organizational structures, these scholars argue, arise from institutional constraints imposed by states and professions. When a measure has been successful in a hegemonic organization, that measure quickly acquires legitimacy. Other organizations in the field now face coercive, memetic, and normative pressures to adopt the same measure, even when there is little "objective" reason to do so. This organizational isomorphism has little to do with efficiency and a lot to do with legitimacy and competition in uncertain environments where it is unclear which measures will be most successful.[15] In short, a form—be it a legal structure, an idea, a set of practices, or some combination of these—spreads because of its legitimacy and is then incorporated into rules, professional practices, and cultural norms. The result is that the form is institutionalized—that is, it becomes the taken-for-granted, commonsensical way of operating.

Because the adoption of a form is driven by perceived legitimacy needs, not the practical needs of the local situation, new institutional theory has developed an important corollary. This corollary distinguishes a formally adopted structure from actual organizational practices; Meyer and Rowan argue that when an organization adopts a formal structure, this adoption is often symbolic and has the status of a myth. Organizations can then maintain internal coordination by decoupling their actual practices from the myth. This decoupling "enables organizations to maintain standardized, legitimating, formal structures while their activities vary in response to practical considerations."[16]

New institutional scholars theorize that inconsistencies between the form and actual practices are resolved internally by decoupling. Globalization scholars have identified an additional solution: they find that when one country adopts another's policy, the adopting country often resolves the discrepancy between the new policy and the practical needs of the local situation by publicly creating its own variation of the adopted policy. Globalization, the literature suggests, creates locally operational and locally meaningful variations on the global theme. The global form has to "make sense" locally, and that means negotiating its details to create a locally viable version of the global theme. As the global form is modified, its meaning is also transformed through local actors' interpretive work.[17] Sometimes, such changes end up having a recursive influence that modifies the original global theme.[18] Taken together, the sociological scholarship on how a global form spreads and gets adopted shows that global diffusion creates a surface-level homogeneity accompanied by detail-level differences in local implementation.

In the case of science and technology policies that promote a more entrepreneurial academia, the enviable burst of the biotechnology industry in the United States—which happened simultaneously with the enactment of the new policies—inspired other countries to adopt similar policies. Japan is one of the more than twenty countries that adopted the U.S. structure for commercializing university research over the past three decades; others include Brazil, China, Denmark, Finland, Germany, Italy, Malaysia, Norway, the Philippines, Russia, Singapore, South Africa, South Korea, the United Kingdom, and more. The U.S. commercialization structure that these countries sought to adopt is characterized by university ownership of intellectual property, strong intellectual property rights, university technology transfer offices that maintain strong control over the university's intellectual property, and the general promotion of commercialization both by universities and by science and technology policy more broadly. The Japanese government started to see its own science and technology policies as suboptimal in the mid-1990s and by the early 2000s had crafted a model that imitated the policies of the United States. The imitation was so conscious and explicit that the act that allowed national universities to hold intellectual property rights to faculty inventions was explicitly nicknamed "the Japanese Bayh-Dole Act."

Since Japan so consciously imitated the U.S. model of research commercialization, should we assume that Japanese academia and its structure for commercializing research are now identical to the U.S. model? More abstractly, does the local imitation of a global form generally result in the replication of the original? The new institutional and globalization scholars reviewed here would answer with a firm no, and indeed, despite its best efforts, Japan didn't completely replicate the United States' rules and practices for commercializing faculty research. What the aforementioned theories can't elaborate, however, is *how* a global form "settles in"—except for correctly pointing out that there will be decoupling between the formal rules and actual practices and that a variation will emerge. But a variation does not "emerge" of its own accord; people create it. How do they do so? What are the processes through which the introduction of a global form reshapes institutions in a particular local context? How do the people affected by the introduction of a global form respond to it and shape its adaptation to the local environment?

The Japanese university scientists I interviewed in this book were among the institutional actors who went through the changes and created new institutional practices. Much of organization theory assumes that institutional actors are passive recipients of change who can only respond to the adoption of new global policies by first resisting and then adopting a compromised version of them. By contrast, my analysis shows how the Japanese university scientists actively interpreted their new environment, worked with the new institutional landscapes, strategized to create new ways of commercializing their inventions, and made the new policy theirs by claiming it as a "Japanese way" to commercialize inventions. In doing these things, they altered how the new rules were enacted. By foregrounding the institutional actors who inhabit and accomplish institutional change, this book advances the microfoundations of institutional theory.[19]

New institutionalism's relatively thin coverage of its microfoundations—what actors actually do—has its roots in the theory's development. What we call "old" institutionalism—pre-1960s institutional theories of organization—explained organizational stability and change as the results of intraorganizational politics, interorganizational competition over resources, and the purposive actions of various organizational actors.[20] "Old" institutionalism did not pay much theoretical attention to the larger societal environment in which an organization must thrive. The core breakthrough of new institutional theory was to shift the focus away from intraorganizational struggles and see the larger picture, in which organizations turn out to be surprisingly and increasingly similar. New institutionalists attribute this homogeneity to organizations' pursuit of legitimacy and survival within the organizational field. As the analytic focus shifted from intraorganizational negotiations to widespread, taken-for-granted schema, new institutionalists encouraged empirical studies that focused on the global diffusion of structure instead of the internal struggles of an organization.[21]

New institutional theory has undergirded many important studies of organizations and global diffusion of a norm; it has also been used to develop theories of world society that consider why countries are becoming more structurally similar. But as new institutionalism's prominence rose, so did its discontents. Its focus on macro-level structure, critics argued, skated over what happens inside organizations, leaving the relationship between action and structure severely undertheorized.[22] Consequently, some institutional scholars shifted their theoretical focus in the early 2000s from global diffusion to the inner workings of institutionalization processes in an effort to revive theoretical attention to the "enaction, interpretation, translation, and meaning" that characterize individual actions in concrete social situations.[23] The individual actions then culminate in the institutionalization of the new norm in local spaces.

This book advances that effort by further developing what scholars have termed "inhabited institutionalism"—an empirical and theoretical corrective to new institutional theory's macro-level focus that aims to reconnect the actions and practices of institutional actors with the dynamics of larger-scale institutional change.[24] Inhabited institutionalism brings people back into theories of organizational change and stability and theories of the relationship between structure and action. Intellectually rooted in the symbolic interactionist ethnography of Herbert Blumer, Everett Hughes, and Anselm L. Strauss, inhabited institutionalism aims to understand how institutional actors interpret and change their environment and identify the ground-level mechanisms by which organizations respond to environments. The focus of analysis moves from the external pressure that makes organizations homogeneous to the ways institutional actors negotiate new practices when a new form is adopted. Institutional inhabitants work out the new form—the "variation"—by translating institutional pressures into concrete rules, objects, and actions; by trying to solve pragmatic problems;[25] and by creating new meanings as they continue to conduct their organizational lives.[26] What emerges from this line of work is a much more inventive actor: actors creatively use local and extralocal meaning systems and material resources to negotiate new sets of practices.[27]

For example, drawing on organizational ethnography, Amy Binder showed that, despite increasing pressure to follow funding-related bureaucratic rules, actors in a transitional housing agency creatively used multiple institutional logics—systems of cultural elements groups and organizations use to organize their activities in time and space—to arrive at decisions they considered professional, compassionate, and accountable.[28] The actors' interactions, skill sets, commitments, and proximity to different environmental pressures shaped the organization's eventual action. The creative use of cultural scripts may extend beyond the original organization, for example, when state attorneys employ the logic of rehabilitation typically associated with clinicians.[29] In sum, these studies shed light on the creativity of actors under institutional constraints. Inhabited institutionalism invites scholars to fully embrace institutional life as an object of study and to approach each institution as one of the many possible configurations of institutional myths, organizational systems, individuals, and actions.[30] By connecting the micro-level actions in an organization with the macro-institutional norm, inhabited institutionalism is able to explain the mechanisms of how an institution changes, why variations emerge during institutionalization, and how a new form settles down—or doesn't.[31]

While inhabited institutionalism has created a good road map for understanding the local actor's role in changing institutions, scholars in this tradition haven't yet looked at the institutionalization of a globally diffusing form—something new institutionalism originally set out to explain. *Cosmopolitan Scientists* helps fill this gap by examining how Japanese university scientists created and institutionalized their own version of a global form. I analyze government and university documents, patent records, and in-depth interviews with Japanese university bioscientists to identify three key themes in local actors' renegotiations of university-industry interactions. First, the previous gift-exchange-like practices exerted a *pulling effect* on how the new practices were created and institutionalized; actors shaped the new practices to resemble the old ones. Second, the scientists were *institutional travelers*—professionals who inhabited more than one institution, in this case, Japanese and U.S. academic institutions. Their dual belonging to Japanese and U.S. academia shaped how these scientists devised their commercialization strategies. Third, *nationalizing accounts*—accounts that frame local conditions, actions, and actors as national ones—permeated the scientists' talk about their commercialization practices after the policy change. They successfully labeled the new practices as "Japanese"—as opposed to foreign, global, or American. The new policy had been adopted from the U.S. Both despite and because of this heritage, the scientists used loose coupling—the gaps between the rules and the way they were enacted—as a resource for everyday nationalism. I refer to these three themes as (1) *the resilience of local institutions and its effect on loose coupling*, (2) *the creative adoption of rules by institutional travelers*, and (3) *the creation and maintenance of nationalizing accounts to enable scientists and their practices to remain "Japanese."* These three themes were essential to the institutionalization of global commercialization policy in the Japanese context.

Fundamentally, this book shows that enacting a global policy in a local space requires the localization of the relevant institution and the actors who identify with it. As new institutional theory suggests, a decoupling between formal rules and local contingencies is a hallmark of any institution. By examining how local actors adapt a global form, the book shows how institutional actors inhabit that decoupling and why: to reconcile the global with "the way things used to be," to exercise creativity as a cosmopolitan local actor, and to incorporate the global into the actor's identity. By looking at these processes at the micro level, this book captures the work local actors do to accommodate the global form and eventually make it *local*. Tying together micro-level actions, interactions, and practices with macro-level policy diffusion highlights the variety of relationships between these two levels. We see tensions and contradistinctions between local practices and global rules, local actors who are at the same time institutional travelers, and variations on a global theme as quintessentially local accomplishments. It is in these tensions and fissures that creative action resides.

Cosmopolitan Scientists considers Japanese scientists' response to this change in science and technology policy as a case study of globalization and its effects. In doing so, the analysis also touches on questions of economization in contemporary society. As Elizabeth Pop Berman and others show, the rise of commercialization of research in the United States and around the world is just one example of the general domination of a neoliberalism that defines and assesses policies by their direct effect on economic matters.[32] Seventy years have passed since the founder of the sociology of science, Robert Merton, described universalism, communalism, disinterestedness, and organized skepticism as the organizing norms of science.[33] In the twenty-first century, science revolves around commercializable innovation. Readers interested in the moral implications of science may ask whether this commercialization and the consequent change in the nature of academia are good or bad. What is lost, we might ask, when the merits of science are measured by their economic impact?

Many books have praised academia's enhanced potential to create innovative products (especially medications that save lives), criticized the corruption and fall of academia at the hands of financial interests, or tried to steer a course between these two extremes.[34] By contrast, the primary goal of *Cosmopolitan Scientists* is to understand how Japanese scientists reshaped their practices of commercialization. For the most part, I bracket the question of whether the reorganization of science to serve the interests of commercialization is good or bad. My findings do, however, offer insight into how scientists have responded to the new entanglements between science and economy. The sociologist Viviana Zelizer—along with others partaking in the "cultural turn" in economic sociology—suggests that "spheres of life" are much more interconnected than we often assume they are. Instead of assuming that intimacy and payment for sex cannot coexist, for example, Zelizer points out that they do coexist and that people work to create practices and meanings that sustain the coordination of these different goods. Our analytic focus, Zelizer argues, should move away from presumed conflicts between spheres of life and toward the delineation of actors' relation work—that is, the ways people negotiate and create acceptable combinations of social relations, transactions, media of exchange, and meaning.[35] From this perspective, the question of science's moral quality can be rephrased as "What are the negative implications of the new ways scientists relate to the act of commercialization?"

Looking at the history of Japanese academia, it is easy to see the deep entanglements between science and technology and modern nation formation. Japanese academia has been concerned with economic matters since its inception. Established as a part of rapid modernization projects after the (re) opening of the country in 1858, Japanese universities have always juggled the intertwined missions of education, knowledge production, and contribution to economic development. So we cannot understand Japan's shift to more commercially oriented academia as a story of an old pure scientific model giving way to a new twenty-first-century neoliberal entrepreneurialism. Instead, we have to look at how Japanese university scientists in the midst of institutional change have renegotiated their relationships with firms and universities. What is the "worth," we must ask, of doing science in this post-1990s period of economization?[36]

Japanese university scientists manage their exchanges and relationships within the options available to them. As the social theorist Steven Lukes points out, the ability to lay out the possible options, along with possible repercussions that actors must expect, is a form of power—the power of policies that increasingly emphasize the value of innovation and economic rewards.[37] As this book shows, Japanese university scientists creatively reshaped their commercialization practices and their scientific lives after the Japanese Bayh-Dole Act and related policies were instituted. Studies of the economization of academia, which tend to focus on the U.S., have pointed out that the deeper hybridization of academia and industry may cause some negative impacts on academia, such as the brain drain from academia to the commercial firms,[38] disregard of basic science,[39] and narrowing of research topics.[40] Studying Japanese academia brings to light another point, perhaps so obvious as to be taken for granted in the U.S. case. What the new policy brought to universities was money—a lot more of it—and contracts. Seen in this light, the economization of academia is not the hybridization of two spheres; it is the encroaching of the economic sphere, which has to be seen in a more critical light. What disappeared from academia in the U.S. and is disappearing in Japan are scientists who take the stance that scientific knowledge production and money are irreconcilable.

In closing this introduction, then, it is quite apt to return to the case of Dr. Honjo and Ono Pharmaceutical. In 2021, many years after the beginning of their legal battles, Dr. Honjo and Ono Pharmaceutical reached an agreement that Ono would pay five billion yen to Dr. Honjo as a settlement and establish the twenty-three billion yen Ono Pharmaceutical and Honjo memorial research fund at Kyoto University.[41] But beneath the surface of this high-profile and contested case are thousands of negotiations—formal and informal—that permeate the world of university research commercialization. These negotiations are not only about deciding the price of research collaboration, innovation, patents, and pharmaceuticals arising from academic research in Japan. They are also, as this book shows, about trust, personal networks, working with changing rules, individual strategizing, and what it means to be Japanese and a university academic. Pricing these less tangible goods is no easy job, but for Japanese university scientists, it's part of their profession.

Outline of the Book

Chapter 2 establishes the empirical ground for the book's analysis. It describes the emergence and institutionalization of a more commercial academia in the United States and the global diffusion of this model during the 1990s. The modern university has never been an ivory tower dedicated to pure science;

it began as an elite educational center before evolving into an organization for research and mass education. This chapter traces these changes in higher education policy and the organizational structure of universities in the United States, with a particular focus on the rise of commercialization. The chapter then describes the corresponding history of Japanese academia, from the establishment of Imperial Universities during the Meiji period of rapid modernization to the adoption and implementation of policies in the 1990s and 2000s that closely imitated U.S. policies designed to promote the commercialization of university research.

Chapter 3 describes how Japanese university scientists practiced science before the formal policy change. In that era, scientists developed trust-based relationships with firms, and industry ties were an integral part of scientists' management of education, their labs, and their research. Before the 1980s, funding and resources were scarce in Japanese academia. University scientists often welcomed firms' requests to collaborate because the collaboration brought funding in the form of donations, ideas, and free labor. Scientists and firms established gift-exchange-like relationships that enabled collaboration without legal contracts, formal compensation, or specification of ownership.

Chapter 4 moves to the present to show how Japanese university bioscientists have reshaped their scientific and commercialization practices since the formal policy changes were instituted in the late 1990s. Specifically, the old gift-exchange-like practices exerted a "pull" that affected the shape of the new practices. Even when the scientists followed the new rules, they created practices that were effectively quite similar to previous ones. For example, they continued to give their intellectual property rights to collaborating firms, often negotiating with university administrators so that the collaborating firm and the university could co-own their intellectual property—a very unlikely arrangement in the U.S. The chapter shows how scientists used their agency to refract the new rules through the lens of established practices to create a uniquely "Japanese" version of the new entrepreneurial academia.

Chapter 5 complicates the picture presented until that point by introducing Japanese scientists who were *institutional travelers*—who had inhabited the U.S. commercialization world in the 1980s and 1990s. Postdoctoral training in the U.S. or Europe was a popular career path for elite Japanese university scientists, and as a result, many top scientists had lived in the United States when their American colleagues were actively engaging with industry. This firsthand exposure influenced how these scientists shaped their commercialization practices after Japan's policy shift in the late 1990s. Many of them didn't welcome the policy changes, because they had already incorporated American-style commercialization practices into the Japanese structure before the policy environment shifted. These scientists had become accustomed to using a combination of Japanese informal practices and U.S.-style formal practices, and many had commercialized their inventions outside of Japan through their international connections. As these scientists' stories make clear, the field of university research commercialization in Japan is a juxtaposition of multiple systems of interaction between scientists and firms.

Chapter 6 shifts gears to explore the discourse of Japanese cultural specificity that these bioscientists used to make sense of their commercialization practices. When these scientists compared their more trust-based, less formalized practices to their U.S. counterparts, they often attributed the differences to Japanese culture. University and government policy documents reinforced this notion of a "Japanese way of interacting with industry" by repeatedly using it to explain how Japan's commercialization practices have formed and reformed. In reality, trust-based, informal relationships between academics and industry partners have been observed in many countries—even in the U.S. before 1960. But by describing such relationships as characteristically Japanese, actors justify and reinforce both Japan's emergent system for commercializing university research and the Japanese people's belief in Japan's "specialness" despite the country's continuous imitation of foreign models.

Lastly, chapter 7 summarizes these findings and discusses their empirical and theoretical significance. Empirically, the book shows how the adoption of the global model of research commercialization reshaped Japanese academia. As the preceding chapters make clear, to say that Japanese university scientists decoupled the formal rules from actual, emerging practices is somewhat of an understatement. In fact, their responses were complex and creative and resulted in a form of commercialization practices that is little like either the old ones or the U.S. ones. Their responses created a new Japanese academia that juxtaposes different paths to commercialization: previously established trust-based practices coexist side by side with faculty startups based on university-owned intellectual property rights, while university-firm co-patenting has taken root as the de facto standard for patent ownership. Scientists who are institutional travelers act as entrepreneurs in both the U.S. and Japan while keeping their university jobs in Japan. On the discursive level, Japanese university scientists and government actors engage with nationalizing accounts that attribute the differences between the global rules and what transpired in Japan to "Japanese-ness."

Global diffusion of policies creates local adaptations that are rooted in both local institutions and their cosmopolitan professionals' multiple belongings. These policies have to be localized to the extent that local actors feel ownership over the processes the policies address. In that sense, the localization of a global form creates an opportunity to enact national identity; in this case, professionals contributed to a discourse that posits an imagined community that is distinct and special precisely as the world becomes more globalized. What is locally institutionalized after the diffusion of a policy, then, is often a very local accomplishment, as local actors determine both what is practiced and how the new policies and practices are understood.

TWO Commercialization of University Research in the U.S. and Japan

The Purpose of the Japanese Bayh-Dole Act:

1. To promote research activities on technology
2. To promote efficient business activities based on such research result

OFFICE OF THE CABINET, JAPAN 2015[1]

THE *JAPANESE BAYH-DOLE ACT* started as a clause in *Special Measures for Industrial Rejuvenation* in 1999, later becoming part of a permanent law called the *Industrial Technology Enhancement Act*. But since its inception, it has always been called by its nickname, the "Japanese Bayh-Dole" Act (*nihon-ban bai-do-ru hou*). This nickname is derived from the U.S. Bayh-Dole Act enacted in 1980 under the sponsorship of Senators Bayh and Dole. When the Japanese university scientists I spoke with talked about Japan's commercialization policies, they referred to them collectively as "the Japanese Bayh-Dole Act." Senators Bayh and Dole probably didn't expect that their names would become so well known in Japan.

This nickname and the ubiquity of its use indicate how commonsensical and intentional policy imitation is in Japan. Japan's modern history has been shaped by imitation of the technologies and institutions of the West since it reopened to foreign powers in 1858.[2] In the early twentieth century, Japan achieved rapid modernization and industrialization by swiftly learning and importing economic, political, and legal systems as well as science and technology from Germany, France, Britain, and the United States. Its ability to imitate is said to be one of the main reasons its economy recovered so dramatically after World War II; this recovery has been described as an "economic miracle" brought about by "the culture of copying" in Japanese manufacturing.[3]

But imitating global policies is far from a Japanese idiosyncrasy—many countries copy hegemonic rules, although perhaps not as conspicuously as Japan. Globalization has made the world an increasingly interconnected web of political, economic, and legal systems, and as globalization continues, different countries' systems are becoming more similar than ever before. Global standards and legal frameworks created in powerful countries are quickly adopted by others;[4] ideas such as human rights, gender equality, and diversity spread throughout the world, reshaping local spaces in not only their formal structures but also their cultural ideas.[5] Imitation and adherence to global standards remain effective strategies for less powerful countries—if not for those standards' inherent superiority, then for the benefits of legitimacy and international compatibility they confer. The global diffusion of policies is thus accompanied by a process of localization as foreign systems are translated and adapted for local contexts.[6]

Global diffusion of policies works differently depending on the nature of the associated norm, the local environment, and the difference between the norm's original environment and its new local environment.[7] As an example, the multilateral treaty on gender equality—the Convention on the Elimination of All Forms of Discrimination against Women in 1979—has been ratified by 189 countries, but it probably did not affect Libya the same way it affected New Zealand, especially since the treaty encompasses vast areas of sociopolitical life and only requires ratifying countries to make a "reasonable effort" to adhere to its terms. So to understand how policy imitation led to increased commercialization of research in Japan and the local consequences of that process, we need to understand both the policy's original form and characteristics and the local situation into which it was adopted, including the possibility of translation. In other words, we need to understand the environments of both the diffusing institution and the receiving institution.

In considering the diffusing institution, we need to understand the origin of the shift in academia that led to the original Bayh-Dole Act, including how neoliberal policymaking, the birth of biotechnology, and changes in patent laws created a new, more entrepreneurial academia in the United States. In considering the recipient, we need to understand the situation before the disruption brought about by the policy imitation, including how the relationship between academia and industry had previously been managed in Japan, what that relationship had meant to the parties, and what legal and institutional scaffoldings for commercialization had supported that relationship.

Only by looking at both sides of the process can we understand how the adoption of U.S. institutions of commercialization interrupted Japan's previous university-firm relationships and what material and interpretive resources scientists had available for the tasks of deciphering the changes and creating solutions. The institutional change in the U.S. was a cultural as well as a legal shift—in other words, the new policies and rules presupposed certain ways of thinking about commercialization.[8] When brought to a new context, such presuppositions may not work and may instead spur local actors to negotiate new ways of thinking about and interacting with industry. Additionally, in the U.S. case, the rise of commercialization and the associated policy framework evolved simultaneously during the 1970s and '80s; by contrast, when Japan adopted the new policies, they had to be reconciled with a long-established and very different set of practices and ways of thinking about university-firm interactions. In such a situation, how new policies are accepted and modified depends heavily on local actors' previous practices and the meanings attached to those practices.

This chapter depicts the historical and institutional context in which Japanese university scientists adopted and adapted the commercialization rules Japan imported from the U.S. I present recent statistics that describe university-firm interactions, university patenting, and entrepreneurial activities among Japanese university scientists following the policy change. Official statistics from the Japanese government show a linear increase in all targeted activities, including the number of collaborative research projects, the total amount of research funds universities received from firms, the number of patents that had universities as assignees, and the number of faculty-involved startups. But the official statistics explain only a part of the picture; they fail to capture two crucial activities. The first is the many informal interactions between university scientists and industries in Japan—informal interactions that are still common after their official "retirement" by the Japanese Bayh-Dole Act. These informal interactions include donation-based research collaborations (*shogaku-kifu*, "donations for research support") and professor-involved patents that do not have universities as assignees. Both of these are still very much part of the academic lives of Japanese university bioscientists. Second, because the professional scientific community is strongly international, Japanese university scientists sometimes work with firms outside Japan and commercialize their research outside Japan. Some of these activities are visible in the official statistics, which record patenting behavior in the U.S. and collaborative research contracts with foreign firms. But other activities, such as serving as a scientific advisor for a startup in the U.S., are very hard to find out about without directly talking with the professors themselves. The role of the official statistics, then, is to prepare us to dive into the interviews, which explore more thoroughly what university scientists' practices of research commercialization look like.

Commercialization of University Research in the United States

Commercialization of university research became the U.S. university's third mission, after research and education, thanks to three interrelated events: the public success of early biotechnology patents, the Bayh-Dole Act's provision that universities can own the patents to their professors' inventions, and a general policy shift toward defining the value of science by its ability to promote innovation and contribute to the national economy. The combination of these three events legitimized and institutionalized the commercialization of university research.

University research had begun to be commercialized as early as the 1950s, but only as isolated, relatively rare instances associated with select universities. There was little institutional scaffolding or incentive for professors to invent until the 1970s. Before that time, science was regarded as a public enterprise, and any discovery emerging from university research was thought to be public. University policies discouraged the patenting of inventions derived from university research, and scientists who did pursue patenting had to surmount a fair number of bureaucratic hurdles before intellectual property rights could be assigned to anyone other than the funding organization such as the National Science Foundation (NSF) or National Institutes of Health (NIH).[9] Each funding organization had a different set of regulations dictating how to handle patents that arose from their funded research, which meant that each funding agency had to resolve patent ownership issues on a case-by-case basis if a professor pursued intellectual property rights.

This part of the bureaucracy wasn't streamlined until the Department of Health, Education, and Welfare introduced institutional patent agreements (IPAs) in 1968. IPAs were agreements between a funding agency and a university that allowed grantee nonprofit institutions to obtain patents for inventions arising from federally funded research. IPAs gradually became more common over the course of the 1970s, and by the end of the decade, the majority of research universities had IPAs with each of the major funding institutions.[10] Still, the rule of thumb was to publish but not patent. Not many scientists were interested in patenting at that time, and some were strongly against it. Some entrepreneurial professors and universities took advantage of IPAs, and some universities that were particularly interested in commercialization, such as the University of Wisconsin–Madison and Stanford, created offices for the purpose of commercializing their professors' discoveries. But in general, owning and profiting from intellectual property remained out of the scope of what an average university scientist would consider.[11]

That all changed in the 1980s, and university patenting became a common affair in research universities by the 1990s. In 1980, about 350 patents were granted to U.S. universities; in 1992, it swelled to 1,400.[12] In 2004, the number grew bigger to about 3,300.[13] In 2012, about 4,800.[14] The institutionalization of university patenting and faculty entrepreneurship has attracted scholarly attention due to its swift rise; within twenty years, faculty patenting shifted from being something of a disgrace to becoming an expectation.[15] What made such a dramatic change possible? Much of the explanation lies in early successes at elite institutions—namely, Stanford, MIT, and the University of California–San Francisco (UCSF)—and the mutually reinforcing phenomena of increasing numbers of professors becoming interested in commercialization, university administrators promoting commercialization, and university administrators and policymakers creating institutional scaffolding to support commercialization. Early biotechnology patents were hugely successful—they were not only medical breakthroughs but also extremely lucrative to the associated universities, professors, and companies.

In the early 1970s, Stanley N. Cohen of Stanford University and Herbert W. Boyer of UCSF developed a laboratory process for transplanting and recombining DNA from different species. Their publication, "Construction of Biologically Functional Bacterial Plasmids in Vitro," signaled the beginning of genetic engineering—along with its commercial applications.[16] Stanford's Technology Licensing Office—one of the few university-based technology transfer offices at that time—filed to patent Cohen and Boyer's scientific procedure in 1974, and the patent was granted in 1980. Without the intervention of the technology office, the two scientists would have simply published the results in science journals without seeking IP protection. The procedure turned out to have immense commercial potential. In 1976, Boyer and venture capitalist Robert Swanson formed a startup company called Genentech that produced recombinant human pharmaceuticals; Genentech began producing human insulin in 1982 and human growth hormone in 1985.

This single instance, led by just a couple of scientists and their elite university, had a major impact on how commercialization of research was institutionalized in the U.S. in the 1980s and '90s. And by the time the U.S. Patent Office granted the first patent associated with this research, for the procedure of recombining DNA from different species, the direction the commercialization of university research would take in the U.S. had been determined. The discovery and its commercial success demonstrated the financial potential of university research and ignited policy discussion. And the specific way the invention was protected—by a university-owned patent and a commercial license to a biotechnology startup funded by a venture capitalist—served as a blueprint for how universities could take financial advantage of faculty research.

A regulatory framework for biotechnology was quickly established while these early successes were happening. In 1980, the U.S. Supreme Court's *Diamond v. Chakrabarty* decision ruled that living organisms engineered by man were potentially patentable under existing statutes. In response, the U.S. Patent Office began issuing property rights on all manner of biotechnology-altered living organisms and their components. Commercial biotechnology had been born.

These events were the precursors of the new academia, and the consequent institutional and legal changes cemented this transformation. The reform of intellectual property rights gathered momentum when Congress passed the Bayh-Dole Act (formally, the Patent and Trademark Law Amendments Act) in 1980. Championed by senators Birch Bayh and Bob Dole, the law provided a single and simplified mechanism for universities (and other nonprofits and small businesses) to own intellectual property rights even if the invention was based on federally funded research. The act enabled universities to obtain IP rights to faculty inventions without establishing an umbrella Institutional Patent Agreement, which radically streamlined the process by which institutions could claim ownership of professors' inventions.

These changes to intellectual property protection were not isolated instances, nor was the timing of the Cohen-Boyer patent a coincidence. The broader trend was the creation of a more commercially oriented, entrepreneurial university, and this trend was reinforced by the development of these practices and regulations. The prospect of biotechnology and the initiative taken by Stanford were both highly visible to legislators throughout the 1970s and '80s, and this visibility propelled changes in federal regulations that benefited universities and changed how legislators thought about the value of university research. Legislators saw the commercial potential of biotechnology, which turned out to generate very effective pharmaceutical treatments for extremely common diseases such as diabetes and cancer. The legitimacy of research funding was being scrutinized, but now a new discourse emerged that cast science in the role of boosting the national economy.[17]

In the mid-1970s, when biotechnology was just starting to become a promising venture, mass media and policymakers were attributing the U.S. economic downturn in part to a lack of innovation. The competitiveness of American industry was in doubt and was on the political agenda, and by the end of the decade, policymakers had passed various measures aimed at fostering innovation. Many of these policies contributed to the rise of commercialized academic research. In 1978, the Department of Labor decreed that pension funds could be invested as venture capital, a decision that helped fuel startups created to market university inventions. A large capital gains tax cut that same year also benefited these startups. To regain American competitiveness in the knowledge economy, policymakers also strengthened intellectual property rights throughout the 1980s. These measures were not explicitly designed to reshape universities, but they helped academic entrepreneurs succeed in their ventures and established the reputation and legitimacy of university patenting and entrepreneurship.[18]

During the 1970s and '80s, a new, more commercially oriented academia was constructed through this combination of scientists' and university administrators' pursuit of patents and a policy shift aimed at fostering innovation and economic growth. The scaffoldings for the new entrepreneurial academia and the nation's broader proentrepreneurial shift co-evolved to such an extent that it's difficult to identify the effects of any particular policy such as the Bayh-Dole Act.[19] But the result of this combination of factors was that by the mid-1980s, commercialization and faculty entrepreneurship had become part and parcel of academic life in U.S. research universities. Every single research university in the United States now has a technology transfer office and celebrates its contributions to innovation. One can embrace it or distance oneself from it, but this institutional project has been accomplished.[20]

Roughly ten years after the Bayh-Dole Act was adopted in the U.S., other countries with strong science and technology sectors started to imitate this move. Genentech began producing synthetic human insulin in 1982, and by the end of the 1980s, the market potential of biomedicines had become very clear. European countries and Japan were the first to emulate the U.S. policy environment for commercialization, and other countries followed suit. In 1985, the U.K. abolished the British Technology Group's monopoly over university patent ownership, effectively ending the national ownership of faculty inventions and promoting commercialization through university technology offices.[21] In 2020, the Association of University Technology Managers (AUTM) identified sixteen countries that had adopted an act similar to the Bayh-Dole Act: Brazil, Japan, Russia, China, Malaysia, Singapore, Denmark, Mexico, South Africa, Finland, Norway, South Korea, Germany, the Philippines, the United Kingdom, and Italy. This number becomes much higher when we add countries that have adopted acts that partially resemble the Bayh-Dole Act and countries that do not have a national law governing university patents but largely follow U.S. practices, such as Canada and Israel.

Was the Bayh-Dole Act successful? It's hard to make causal claims about a single strand within a web of changes taking place at the same time; the number of inventions is likely to have risen even if there were no patent policy changes, and universities in America would likely have become more entrepreneurial even without intervention.[22] But for all practical purposes, the policy was an enviable success. Its success drew the attention of policymakers around the world, and the U.S. structure of university research commercialization started to diffuse across the globe, including countries whose social and economic conditions differ greatly from those of the U.S.[23]

The legitimacy and allure of the new policy were enough for it to travel to Japan, where importing policies from the U.S. is commonsensical. By the late 1990s, Japan was implementing a series of reforms aimed at reshaping university research and commercialization. Japan may resemble the U.S. in its production of scientific and technological knowledge, but its social institutions are very different from those in the U.S. As just one example, most research universities in Japan are national universities, whereas private universities play a large role in academic research in the U.S. To understand the processes and the consequences of the adoption, then, it's necessary to consider the historical context of research commercialization in Japan. How were universities and their research managed in Japan, and how did the introduction of the new policy reshape this environment?

Japanese Academia, 1950–1990

Higher education in Japan has its roots in the country's rapid modernization during the early Meiji period. In 1868, the Meiji Restoration replaced the feudal Tokugawa Shogunate in Edo with a centralized imperial government consisting of politicians from powerful domains—the estates of *Daimyos*.[24] The new Meiji government's mission was to promote industrial and technological advancement to be able to compete with great powers. A modern educational system was a primary focus and was quickly established during the late nineteenth century. In 1877, the University of Tokyo was established by drawing together various teaching institutions of the Tokugawa Shogunate and inviting academics from Britain, France, Germany, and the United States. The Imperial University Order was issued in 1886, and by 1931, nine Imperial Universities had been created in Japan and its territories. Private universities and other professorial training institutions were promoted and governed by other imperial orders, and the number of higher education institutions increased rapidly. In 1895, there were 63 higher education institutions in Japan; by 1925 this number had increased to 257.[25]

The establishment of modern higher education in Japan had nothing to do with the ivory tower ideal and everything to do with industrial growth. From the beginning, universities were established as part of the mission of "Wealth and Military Strength"—one of the two slogans the Meiji government used to articulate its goal of fast-forwarding the transformation of Japan into a modern nation-state.[26] Education policy was unquestionably tied to the national project of a modern Japan. Higher education was tightly controlled, first by imperial ordinances and later by the Ministry of Education, until after World War II. Until 2004, all national universities were directly administered by the Ministry of Education, and their professors were national employees. Private universities slowly prospered alongside the expanding national university system; in 2020, there were 82 national universities, 91 public (for example, prefectural) universities, and 592 private universities.[27]

In the first half of the twentieth century, industries and universities collaborated to a considerable degree in Japan, largely because there was no clear distinction between academia, the government, and industry. For example, companies such as Ajinomoto, Yakult, Toyota, and Toshiba all have their roots in Imperial Universities or the Imperial College of Engineering. Imperial universities' primary mission was to produce human resources that could be hired to serve as government officials, but their secondary mission was to produce highly educated engineers. And because there were virtually no restrictions on professors' business activities, any professors who were so inclined could invent, obtain patents, and collaborate with industry partners. Because imperial university graduates regularly went on to serve in elite positions in the government and military, these institutions became a tight-knit web of former classmates who regularly collaborated with familiar faces.[28]

Imperial Universities' strong ties to industry and government meant that the university was also tied to the military. Imperial Universities, especially engineering departments, had co-evolved with the military, and in the early Meiji period, university engineers lent their technical expertise to the establishment of the modern military. Professors typically held multiple positions in government, Imperial Universities, and industry, and highly ranked naval officers regularly taught in Imperial Universities. Although the relationship between academia and government was close from the beginning, academia remained autonomous until the 1930s, when the relationship between university research and the military started to become coercive. From 1930 to the end of World War II, Japanese academia was controlled by Japan's military government, and research topics and budgets were tailored to fit nationalist military purposes.[29]

It was this symbiotic relationship between universities, government, industry, and the military that made the postwar Japanese government—directed by Allied occupiers—want to distance academia from industry. The Japanese constitution of 1946 dedicated a clause to academic freedom: "Academic freedom is guaranteed." This clause was interpreted to suggest that public employees, including most research university scientists, could not align themselves with particular companies but instead must serve as science and technology experts for the common good. This admonition against commercially oriented academia was further emphasized in the 1960s, when the student movement caught fire in Japan and around the world. A senior professor at one of the best medical schools in Japan recalled,

> When I enrolled in medical school, it was 1961. The university was in turmoil due to the U.S.–Japan security treaty revision in 1960; there was so much political upheaval in universities. Japan was still reflecting on the war, and socialism was still a big influence. [. . .] Industry was seen as a threat to academia, just like what had happened in the industry-university complex

> [during the war]. [. . .] Industry's interest was thought to be a bad thing. This was the 1960s. This trend of opposing cooperation between universities and academia lasted until the 1970s, when the economic miracle started to happen.

Because Japan had no need to streamline mechanisms for commercialization during this period, it lacked dedicated laws that specifically addressed universities in the context of ownership of intellectual property, such as the U.S. Bayh-Dole Act of 1980. National university professors were national employees, so the ownership of intellectual property resulting from research activities in national universities needed careful legal treatment. In principle, all faculty inventions could be interpreted as national inventions, in which case proprietary ownership by private entities would be impossible. As the professor quoted previously notes, though, university-firm interactions became more common during the 1970s, when professors' scientific and technological expertise became a resource for a country experiencing rapid reindustrialization and economic development.

It was in this context that, in 1978, the Ministry of Education published Notification 117, which legally allowed some types of intellectual property based on faculty research to *not* be considered national inventions. The notification cited School Education Law Clause 58, which states, "Universities' objectives are as follows; serve as the center of science, disseminate knowledge, and teach and research specialized knowledge to develop the competency of the public intellectually, morally, and practically." The notification decreed that since technology transfer is not necessarily a part of the objective to "teach and research," professors' inventions were not to be considered employee inventions. The notification instructed all national universities to establish an Inventions Committee and required that all faculty inventions be reported to their university's committee. It assigned ownership to the individual professors themselves, except under certain circumstances. The notification explains this procedure as the best practical solution for commercialization: "If we make all faculty inventions nationally owned when universities and the nation do not have the capacity to acquire and manage inventions promptly and accurately, the result could be that excellent inventions fail to become patents or escape abroad. Therefore, it is more efficient to transfer the rights of faculty inventions to the individual, so that a path to prompt application and potential revenue to further develop the research will be available."[30] Notification 117 was in effect until the enactment of the Japanese Bayh-Dole Act in 1999. It legitimated informal collaborations between university scientists and firms, and because it identified professors as individual owners of their own inventions, it gave professors the leeway to transfer the IP rights for their inventions to their collaborating firms.[31] The notification was thus a de facto governmental endorsement of informal technology transfer. And while Japanese universities had relatively few formal mechanisms for collaborating with firms until the 1980s, there had been an established scheme for receiving donations for research from firms as early as 1964.[32] Called *shogaku-kifu*—translatable as "donations for promotion of research"—such donations are tax exempt and may be allocated to a specific lab, hence a specific professor. Firms commonly made such donations as payment for networking, consulting, or collaborating. During the 1970s and '80s, Japan's Ministry of Education slowly started to implement regulations that enabled more formal interactions such as "sponsored research" and "collaborative research" but never established a clear rule for faculty inventions. In the end, Notification 117 acted as a legal guideline and a legitimate precedent. Overall, during this period, the government did not proactively encourage or intervene in commercialization activities but instead paved the way for university professors to pursue commercialization largely on their own with the help of collaborating firms.

University administrations also lacked formal endorsement, mechanisms, and surveillance of research commercialization during this period. Japanese universities did little, if anything, to involve themselves in commercialization efforts, nor were they given any budget to do that. Until 2004, national universities were vertically integrated as subunits of the Ministry of Education and so had little autonomy to produce internal policies regarding IP rights and commercialization. Private universities also followed the Ministry of Education guidelines. A lack of formal mechanisms and a proliferation of informal exchanges were the de facto standard; most inventions were handled informally between university scientists and firms, and there was no need and no funding to develop university regulations and organizations. After Notification 117 was issued in 1978, universities did establish inventions committees—usually at the departmental level and made up of faculty members—but these committees simply allowed individual professors to own the rights to their own inventions, thus confirming established informal routes to commercialization.

Because there were no formal mechanisms, commercialization of research largely depended on individual university scientists. Until the 1970s, Japanese university scientists generally kept their distance from industry, partly because academia emphasized research and publication rather than application, partly because the laws governing public employees prohibited favoring any particular company, and partly because both the public and university communities generally disapproved of university-industry engagements.[33] The general sentiment against collaborating with industry gradually began to exert less influence, however, as industry needs increasingly propelled industries to connect with professors informally. These informal connections resulted in an increase in the number of university-industry joint research projects during the 1980s and '90s.[34] As such collaborations took root, informal commercialization of research became an everyday affair in Japanese universities; judging from the number of articles co-authored by professors and industry scientists during the 1990s, university-industry collaborations were as common in Japan as they were in the United States.[35]

Without formal mechanisms for technology transfer, and without the formal recognition of patents as part of a professor's scientific work, commercialization through patenting remained the job of private firms. Often, a patent arising from collaboration between a firm and a professor listed a university scientist as an "inventor" and the collaborating firm as the "assignee"—the owner. Legally, the owner—the assignee—holds the property rights to the invention, so firms were usually happy to follow this arrangement. As a courtesy, the firm was expected to do favors for the professor who gave the firm the rights to the invention. A professor who was active through that period summarized this informal mechanism of technology transfer during an interview. Speaking of industry collaboration in the '80s and '90s, he said, "At that time, there were no such mechanisms, so researchers had two choices: either pay out of pocket to be an inventor and an applicant (assignee) or just let the firms take over. I chose the latter. So I have patents that I invented but that I never filed myself. The firm would pay all the costs for the patent and for the technology as well. In return, we'd receive donations, increased donations. That was the exchange." This professor's description exemplifies a typical pattern of technology transfer before the institutional changes of the 1990s and 2000s. As in the example of Dr. Koyama in chapter 1, the encounter between the firm and the university professor was considered a personal relationship more than an organizational arrangement. Typically, the firm scientists who directed a firm's research would contact a professor whose research might benefit the firm. In many cases, they already had personal connections; for example, they might have worked in the same lab at some point in their careers, the firm scientist might have been a former student of the professor, or they might have met at academic conferences. Rough ideas about the collaboration and its budget, personnel, and purpose were written in an informal document called an *oboe-gaki*—roughly translated as "memorandum"—which served as a guidepost.[36] The legal implications of *oboe-gaki* are unclear; it was rare for either party to try to bring it to court.

Informal collaborations funded by *shogaku-kifu* donations usually took place in the professor's lab; in some cases, the firm sent their skilled researchers to the professor's lab to provide research assistance anywhere from a few times a week to full time. Because it was also common for firms to send their researchers—who often only had MS degrees—to train for a PhD while remaining in the firm, these arrangements worked seamlessly to continue the relationship between a university lab and a firm. The professors often appreciated the extra help the firm scientists brought to the lab; they were often well-trained and willing to help not only with research but also with teaching young lab members. Because technology transfer was not recognized as part of a university scientist's job, scientists often simply passed the rights to their inventions on to collaborating firms in exchange for increased donations later on. Typically, firms acquired the sole ownership of IP rights to university-based inventions.

In summary, Japan's government and universities decreed that commercialization was not part of the university's mission. The resulting lack of formal mechanisms for commercialization had the counterintuitive effect of enabling university scientists and firms to develop their own practices of collaboration and technology transfer. Three practical ramifications of this system affected how the later formalization of research commercialization took place. First, the tradition of universities taking a back seat in commercialization activities shaped both universities' and scientists' expectations. Second, the tradition of informal collaboration created informal trust relationships that scientists and firms often appreciated and were reluctant to sacrifice in favor of the new formal system.[37] Third, the combination of a lack of institutional scaffolding, the relatively small Japanese market, and a lack of dedicated resources led many renowned scientists to commercialize their inventions in the U.S. or Europe.

Legal and Institutional Changes, 1995–2005

In the 1990s, Japanese government officials increasingly began to see the Japanese system of research commercialization as suboptimal. This perception was in great part caused by comparing Japanese academic entrepreneurship with that of the United States—especially in biotechnology, in which commercial applications of university-originated inventions had created a whole new industry that had boosted the U.S. economy. Looking at the greener pastures of the U.S. biotechnology industry, the Japanese government moved to promote commercialization by implementing new policies.

These new policies were developed in a broader context of ongoing science policy reform. In 1995, the Basic Act on Science and Technology was formulated as the first umbrella policy to govern science and technology. The act asks the Council for Science, Technology, and Innovation to create a new Basic Plan for Science and Technology every five years to guide the development of more targeted policies and budgets across different ministries. The purpose of the act was to further national development by promoting science, technology, innovation, and the creation and application of intellectual property.

Through the 1990s, the Japanese government implemented a series of related efforts, including relaxing the rules regulating university professors and other public employees, strengthening intellectual property rights, and promoting university entrepreneurship. Intellectual property reforms were designed to render the Japanese patent system compatible with the WTO/TRIPS agreement. In accord with the Science and Technology Basic Law and the Basic Plan, promoting university patents became an official goal, and patent laws were amended to "create a system that allows for strong protection of IP rights to realize a nation built on the platform of scientific and technological creativity."[38] Property rights were strengthened, application fees were lowered, the approval process was streamlined, and extra protections were allowed for drug discovery, among other changes. To address the fact that national university professors were national employees who were prohibited from favoring particular private businesses, the National Public Service Act was amended in 2000 to allow national university professors to work for private firms for technology transfer purposes. In 2004, the National University Corporation Act changed the employment status of national university professors such that they were no longer national employees and instead were governed only by their university's regulations.

The most significant and symbolic legal change affecting university research commercialization is 1999's Japanese Bayh-Dole Act. This act was officially a part of the Act on Special Measures Concerning Industrial Revitalization and was later made permanent by being incorporated into the Industry Technology Enhancement Act in 2000. As the nickname suggests, the Japanese Bayh-Dole Act is an adaptation of the U.S. Bayh-Dole Act; it allowed universities to control IPs arising from nationally funded research and cleared up the legal pathway for universities to capitalize on faculty inventions. Like the original, the Japanese Bayh-Dole Act allowed universities to obtain IP rights even if the invention arose from nationally funded research or research conducted under the auspices of a national university. In 1998, just before the passage of the Japanese Bayh-Dole Act, the government passed the Act to Facilitate Technology Transfer from Universities to the Private Sector, commonly referred to as the TLO Act, to respond to the fact that universities in Japan had no offices or budget for IP and technology transfer. The TLO Act provided government aid to enable the launch of approved technology licensing offices (TLOs). In accord with these changes, both public and private universities gradually started to claim the IP rights to inventions that resulted from their faculty members' research.

The final step in this repositioning of academic entrepreneurship was the National University Corporation Act in 2004. This act gave national universities individual corporation status and allowed them to freely develop their own strategies regarding university-industry collaboration. In response, universities strengthened their industry relations and technology transfer facilities so they could deal with contract research and collaborative research with firms, screening and filing patent applications, and marketing university IPs to firms. As a result of these changes, by the end of 2010, every major Japanese university had a division for university-industry relations, and fifty approved TLOs were working with affiliated or contracted universities.[39]

The policy emphasis on commercialization also became loud and clear, including with regard to research funding. In the early 2000s, applications and reports for grants-in-aid from MEXT and MITI (the equivalents of federal funding organizations such as the NSF and NIH in the U.S.) began to require researchers to report their patenting records.[40] Additionally, application research became increasingly important for securing various kinds of funding, including funds for basic research that might produce applications down the road, funds for startups based on university-led innovation, and matching funds for collaborating with firms.

By 2010, when I began to conduct the interviews for this book, most of the policy and institutional changes had been in operation for more than five years. At that point, in terms of legal and organizational arrangements, Japan probably had an even more favorable environment for faculty entrepreneurship and technology transfer than the United States. At present, Japanese universities are equipped with a new formal system that university scientists and firms are supposed to follow. According to this system, collaborations with industry operate as follows: If a professor and a firm want to conduct a research project together, they sign a formal contract laying out the collaborative research plan that the university's administrators approve. Next, when findings from the research are deemed patentable (and potentially profitable), the professor will report it to the university and the technology transfer office will patent the invention. In accord with the Japanese Bayh-Dole Act, the university will own the invention. If the university and the firm successfully negotiate licensing agreements, a part of the royalty will be paid to the inventing professor and one's laboratory. In addition, if an academic research project conducted without the involvement of industry turns out to be patentable, the professor contacts the university. After screening, the university applies for the patent and, through the technology licensing office, licenses the invention to a firm.

By formalizing the commercialization processes and involving the university in the process, these changes created a procedure for university scientists to follow that eliminated their reliance on firms as well as any ambiguity in relation to responsibility and ownership. The new scheme demands a clear contract between the firm and the university lab and allows university scientists to negotiate the terms of collaboration with the help of university administration and the technology transfer office. In compliance with the Japanese Bayh-Dole Act and other regulations, any inventions arising from such collaborations are owned by the university, and scientists' fair share of the royalties are to be paid upon licensing.

By the 2010s, Japan's adoption of the global, U.S.-originated system of research commercialization had been completed. What have the consequences been of this adoption—and adaptation? Did commercialization practices completely change? As I explained in the last chapter, some loose coupling between the new rules and the actual practices is to be expected. In theory, scientists should have welcomed such changes, because they formally encouraged commercialization and offered clearer terms of collaboration than the informal exchanges. But in practice, the university scientists I interviewed had not replaced their previous practices with the new procedures. Japanese university scientists by and large accepted the new formal approach to university research commercialization—after all, the changes weren't optional. But instead of simply transitioning to the new procedures, they negotiated with them and created a set of new practices that were decoupled from the new formal rules.

The Statistics, the Untold, and Our Road Map

Any official documents from the Japanese government would correctly suggest, based on all the available metrics, that the formal policy change has been a resounding success. The number and size of university-industry collaboration experienced a significant boost after the policy change. As table 1 shows, the number of university-firm collaborations rose during the years of 2006 to 2018. The number of collaborative and sponsored projects[41] between firms and universities rose from 20,674 in 2006 to 37,298 in 2018, and their research budgets rose from about forty-one billion yen in 2006 to about eighty-four billion yen in 2018.[42]

Table 2 then shows the magnitude of firm-sponsored research from 2001 to 2007. Here, there is a clear increase in governmental grants for the national and public universities that far surpasses the inflation rate. Moreover, the other two external sources of funding—"Research Donations from firms" and "Sponsored and contracted research"—both increased during the same period; Thus, more generally, it is clear that the reliance of universities on firms to provide research funding had also increased. Table 2 also shows that the informal donation-based research—research funded by *shogaku-kifu*—was not abolished but rather increased significantly.[43] Although there is little research tracking the increase in research donations, according to Yamaguchi's detailed work on sixty-six major universities during the 2001 to 2007 period, research donations increased by 5.5% per year on average and made up 20–30% of total external funding.[44]

Figure 2.1 and table 3 help to further clarify these changes. As interactions between the universities and firms intensified, the commercialization of university research became more common as well—again, the intended effect of the policy change that hoped to capitalize on university knowledge. Figure

TABLE 1: Number of university-firm collaborations

Year	Collaborative and sponsored projects	Collaborative and sponsored projects with foreign firms
2006	20,674	156
2007	21,802	186
2008	22,927	216
2009	22,973	257
2010	23,610	273
2011	24,073	296
2012	25,095	263
2013	26,571	282
2014	28,037	287
2015	29,981	335
2016	32,356	377
2017	35,066	463
2018	37,298	499

Source: MEXT, NISTEP Science and Technology Indicators 2019. Table created by the author.

TABLE 2: The rise in external funding among sixty-six national and public universities (in 100 million yen)

	2001	2002	2003	2004	2005	2006	2007
Governmental grants	786 (44.6%)	950 (20.8%)	954 (42.6%)	1112 (41.3%)	1132 (38.1%)	1136 (36.4%)	1233 (35.3%)
Research Donations from firms	528 (30.0%)	559 (27.3%)	529 (23.6%)	608 (22.6%)	657 (22.1%)	632 (20.2%)	726 (20.8%)
Sponsored and contracted research with firms	447 (25.4%)	535 (26.1%)	755 (49.7%)	972 (36.1%)	1178 (39.7%)	1354 (43.4%)	1535 (44.0%)

Source: Original tables in Yamaguchi 2010. Table created by the author.

2.1 shows that the number of university-involved startups also increased dramatically from 112 in 1995 to 2,905 in 2020. The official statistics also show a dramatic increase in the number of patents arising from university research. Table 3 shows that the number of university patent applications—the patent applications that have Japanese universities as one of the assignees—rose from 693 in 1998 to 6,373 in 2010.[45] To be clear, these figures only represent patent applications that list technology transfer offices or universities as one of the "assignees"—those who will have the IP rights when the patent application is granted; they leave out projects that involved university professors but did not list a university on the patent application.[46]

What, then, happened to the patents that arose out of informal practices between university professors and firms that were prevalent in the past? Because it was customary for university professors to personally interact with

FIGURE 2.1: University-involved startups

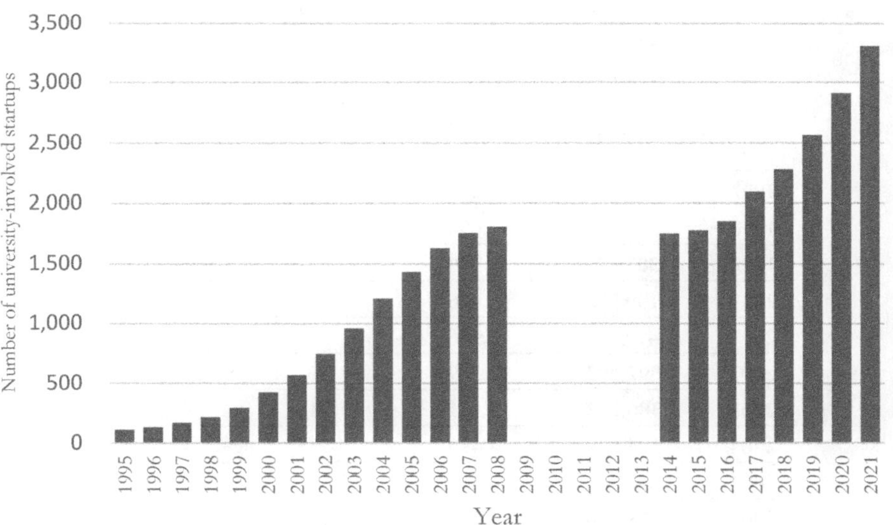

Note: Data for 2009–2013 unavailable in the original. University-involved startups include startups that are based on patents or inventions originated in universities, startups that have collaborations with a university, startups that use university technology transfer schemes, startups that are founded by students and are related to the academic activity, and other kinds of university involvement.

Data from MEXT, NISTEP Science and Technology Indicators 2022.

TABLE 3: Number of university patents (with breakdown for firm co-patents)

Year	Total # of patent applications in Japan	University patents applied for	Co-patents with firms	Ratio of co-patents (%)
1981	215,234	97	35	36.1
1982	234,035	150	58	38.7
1983	251,496	137	49	35.8
1984	281,081	198	102	51.5
1985	297,050	178	82	46.1
1986	311,346	253	134	53.0
1987	329,499	326	230	70.6
1988	328,401	295	201	68.1
1989	337,167	309	193	62.5
1990	354,154	312	223	71.5
1991	354,986	295	220	74.6
1992	355,518	352	257	73.0
1993	349,150	384	277	72.1
1994	336,699	314	225	71.7
1995	351,713	324	215	66.4
1996	359,046	329	233	70.8
1997	372,089	471	305	64.8
1998	383,186	693	417	60.2
1999	383,475	930	473	50.9
2000	409,557	1,416	598	42.2
2001	410,769	1,966	823	41.9
2002	392,493	2,269	931	41.0
2003	386,272	2,931	1,240	42.3
2004	394,039	4,499	1,955	43.5
2005	397,356	6,738	3,006	44.6
2006	378,870	6,962	3,286	47.2
2007	359,342	6,981	3,458	49.5
2008	351,070	6,623	3,640	55.0
2009	315,944	6,304	3,289	52.2
2010	296,225	6,373	3,441	54.0

Source: Table 4-2-16 in MEXT, NISTEP Science and Technology Indicators 2014. Translated by the author.

employees of collaborating firms and give any patent rights to those firms with the expectation of additional donations to the lab, these patents usually list a firm as the assignee and the firm scientists, the university professor, and their team members as inventors. This means that most official statistics—including the ones in table 3—do not include the many patents that list university professors as inventors rather than assignees.

Two data points capture the extent to which university-firm collaborations produced patents before the policy changes. A table MEXT published in 2003 shows that in 1999, of the 1,725 inventions that were reported by university scientists to national universities, only 281 were attributed to the institution—the university or its affiliated TLOs (technology transfer offices).[47] Of those 281 inventions, only 220 made it to the application process; in other words, the universities only acquired patents for 13% of the reported inventions, gave ownership of 84% of reported inventions to the professors who reported them, and decided against pursuing IP rights for 3% of the reported inventions. Once a potential invention was reported to the university but the university did not pursue ownership—87% of the reported inventions in 1999—faculty were free to choose the next step, such as abandoning the idea of patenting, trying to patent by themselves, or giving the patenting rights to firms. Although there is no quantitative data on what happened to these inventions, it is likely that most of the ones that were deemed worthy of securing IP protection went on to be patented through collaborating firms, in which case the firms acquired the rights to the intellectual property. In 2003, when more TLOs were being established and the National University Corporation Act was in the making, 84% of reported inventions (5,781 out of 6,787) were still "sent back" to the professors who invented them. Hence, it is likely that the official statistics on university patents during the 2000s miss a significant number of patents in which university professors were involved but were listed only as inventors. These figures also suggest that firms that collaborated with university professors were actively patenting the resulting inventions even before the policy changes about research commercialization.

The most fine-grained analysis of the increase in patents and their allocation during this period appears in a study by Daisuke Kanama and Kumi Okuwada, who examined three national research universities—Tohoku, Tsukuba, and Hiroshima Universities—between 1993 and 2006.[48] This study examines all patents that listed university professors as inventors, not just those that listed universities as assignees. Figure 2.2, reproduced from this

FIGURE 2.2: Ratio of university-firm co-patents in three research universities

Publication Year	Co-application between university and private firms	Sole application by university-related entities (includes University, TLO, or individual professors)	Sole application by firm	Other (includes patents through JST – Japan Science and Technology Agency or NEDO – New Energy and Industrial Technology Development Organization
Tsukuba U.				
1993~2002	16%	13%	51%	20%
2003~2005	16%	20%	37%	28%
2006	12%	38%	25%	25%
Hiroshima U.				
1993~2002	13%	11%	61%	15%
2003~2005	19%	25%	44%	12%
2006	35%	47%	13%	6%
Tohoku U.				
1993~2002	46%	15%	30%	9%
2003~2005	39%	16%	32%	13%
2006	38%	37%	20%	5%

Source: Kanama and Okuwada 2008. NISTEP Report.

study, shows the ratio of university-firm co-patents in three research universities that Kanama and Okuwada were able to gather detailed longitudinal data about between 1993 and 2006. The authors conclude that the policy changes did increase the number of patents for university professors' inventions, mostly by increasing the number of new inventors. In other words, policies that promoted patenting prompted more professors to patent. However, the study also revealed the continued prominence of firms in patents of university-originated inventions. Even in 2006, "university patents"—patents that are officially recognized and owned by universities—were not the only ways professors patented their inventions. That year, professors at those three national universities still single-handedly gave their IP rights to collaborating firms for about 13–25% of university-originated patents.[49] It is reasonable to expect that the practice of firm ownership of patents has declined since 2006, but given that *shogaku-kifu* research donations have *increased*, not decreased, even after the introduction of more formal policies, it is likely that this practice is still a part of the repertoire of university research commercialization, especially when the university does not have a sufficient budget for IP management. Once a university releases its ownership rights, it is completely legitimate for firms to patent professors' inventions.

The Popularity of University-Firm Patents

Among the officially traceable university patents, one unique arrangement that has gained popularity is the university-firm patent—often referred to as a *kyo-gan* (co-applied) patent. A *kyo-gan* patent has two (or more) assignees, and a firm and a university can use this type of patent to become co-owners of an invention's IP rights. This arrangement is very rare in the United States and relatively rare in Western Europe but has been popular in Japanese academia.[50] As table 3 shows, university-firm co-patent applications made up 40–55% of all university patent applications between 2001 and 2010. This percentage had been even higher before 2000 because prior to the policy changes, intellectual property management was outside the purview of universities, and firms often handled patent application and maintenance. Even after the Japanese Bayh-Dole Act, which enabled universities to be patent assignees, many firms continued to co-own the IP rights to collaborative inventions. This trend seems to be continuing to the present day: as recently as 2019, 63% of all university patents (4,237 out of 6,726) were co-assigned with a firm.[51]

International Commercialization Activities

Science policies travel globally, and the scientific community is international, but national policies are set by nations, and statistics on science and commercialization are accordingly circumscribed by nation. Thus, there is another reason why we don't have good, continuing statistics on patents that arose from university-firm collaborations but are not owned by universities: *methodological nationalism*, the tendency to take the boundaries of the nation-state for granted and use it to construct data and analyses, prevents us from understanding how Japanese university scientists have commercialized their research outside Japan.[52]

However, two statistics from MEXT give us some idea of how Japanese university scientists collaborate with foreign firms and patent their inventions outside of Japan. The first is the number of collaborations between foreign firms and university professors and their associated funding. The breakdown of foreign and domestic firms among firms collaborating with a university in table 1 shows that the number of collaborations between Japanese university professors and foreign firms rose rapidly from 156 in 2006 to 499 in 2018; this period saw a significant jump in funding as well, from 668 million yen to 2.2 billion yen.[53] Officially, formal international collaborations make up only a fraction (about 1%) of total firm-professor collaborations. But because many foreign firms—especially pharmaceutical companies—have incorporated or partnered with Japanese pharmaceutical companies, the reality of foreign collaborations is likely more significant.

The other relevant MEXT statistic is the number of patent applications abroad. A foreign patent application—most commonly with the U.S. or the EPO (European Patent Organization)—directly signals the intent to commercialize. Patent applications abroad are more labor- and cost-intensive than domestic applications, so a university would not apply for a foreign patent without believing the invention stood a good chance of going to market. Table 4 shows the number of foreign patent applications from Japanese universities between 2005 and 2018 and how the invention emerged—that is, whether it arose from a university-firm collaboration, from research donations from a firm, or independently of any firm involvement (note it does not show the IP ownership arrangement). As the table shows, applications increased sharply during this period, from 1,330 to 2,934. The breakdown of firm involvement shows their importance in the acquisition of IP rights, even for foreign patents: from 2011 to 2018, firms were consistently involved in 40–50% of foreign patent applications. Similarly, Robert Kneller shows that 61 out of 125 (49%) U.S. patent applications by Tokyo University had a firm as a co-assignee.[54] This is roughly the same percentage as in Japan, where again roughly half of patent applications have firms as co-assignees. We do not have good quantitative data, however, about Japanese university scientists' informal collaborations with foreign firms through research donations or consulting, their entrepreneurship abroad (such as being a member of a startup in the U.S.), or the licensing of their patents to foreign firms.

Summary and Road Map

The available data on university research commercialization shows that university professors' commercialization activities increased during the late 1990s and early 2000s. The number of university-firm collaborations rose rapidly, as did research funds for these collaborations and the number of university-

TABLE 4: Number of foreign patents (with breakdown for the source of funding)

Year	Total applications (domestic & foreign)	Patent applications—Japan				Patent applications—Overseas			
		Firm funded	Research donations	Other	Total	Firm funded	Research donations	Other	Total
2005	8,527	—	—	—	1,330	—	—	—	—
2006	9,090	—	—	—	1,808	—	—	—	—
2007	9,869	—	—	—	2,987	—	—	—	—
2008	9,435	—	—	—	2,455	—	—	—	—
2009	8,801	—	—	—	2,002	—	—	—	—
2010	8,675	—	—	—	2,185	—	—	—	—
2011	9,124	6,507	2,372	3,907	2,617	1,098	88	1,431	2,617
2012	9,104	6,517	2,323	3,962	2,587	978	67	1,542	2,587
2013	9,303	6,605	2,459	3,915	2,698	1,287	94	1,317	2,698
2014	9,157	6,585	2,355	4,043	2,572	1,151	72	1,349	2,572
2015	8,817	6,437	2,451	3,818	2,380	1,065	80	1,235	2,380
2016	9,388	6,661	2,550	3,891	2,727	1,221	131	1,375	2,727
2017	9,427	6,574	2,545	3,831	2,853	1,323	102	1,428	2,853
2018	9,529	6,595	2,721	3,716	2,934	1,389	79	1,466	2,934

Note: The table shows the funding source for the research the invention is based on; it does not specify the assignees of the patent. The original table is table 5-4-8 in MEXT, NISTEP Science and Technology Indicators 2022. Edited and translated by the author.

The total number of Japanese patent applications from 2005 to 2010 in this table is slightly larger than the numbers indicated in table 3, which was from Science and Technology Indicators 2014. This is because NISTEP updates the additional data and makes corrections each year.

owned patents. After Japan relaxed regulations on faculty entrepreneurship, startups based on faculty inventions also increased steadily.

These statistics tell a simple story: the policy was successful. The reality, however, is more complicated for two reasons. First, as in the case of the U.S., it is nearly impossible—and also somewhat pointless—to try to isolate the effects of the policy changes from the effects of the broader trend toward increased research commercialization. The success and influence of U.S. university research commercialization and the rise of biotechnology would likely have generated increased university-based patent activity in Japan even if none of the associated policy initiatives had taken place. Second, because university-industry interactions were informal before the policy changes, it is hard to know how much collaboration was taking place and how many university-based inventions were patented. Nevertheless, all in all, it is safe to say that universities' commercialization activities increased after the policy changes.

Whereas the general increase in commercialization activities is clear, some details seem to stand out. First, the legacy of informal collaboration between professors and firms seems to have influenced how university professors engage in research commercialization even years after the policy changes. To begin with, the number of patent applications grounded in donation-based research increased after the policy changes. In addition, two different patent arrangements began to coexist after the policy changes: the formally sanctioned arrangement according to which universities own the patents to inventions, and the traditional arrangement in which university professors are listed as inventors and the collaborating firm is listed as the assignee. Second, the ongoing influence of the informal practices extends beyond the firm-professor relationship; it also seems to have given rise to the new popular arrangement in which universities and firms co-own patents as co-assignees. This is a distinctively local arrangement that is rarely seen in the United States.[55] Third, whereas science is global, nation-bounded measurements fail to capture the extent to which Japanese university scientists are influenced by commercialization in the U.S. as well as whether and how they commercialize their research outside of Japan.

The following chapters draw on in-depth interviews to dive into Japanese university professors' experiences of commercialization. Not all of the bioscientists I interviewed were aware of the full history of research commercialization in the U.S. and in Japan. But they were all acutely aware of how the informal, gift-exchange-like practices had worked in university-firm collaborations in Japan; many of them had used these practices themselves for a long time. Under that system, they said, trust mattered, money wasn't an issue, and science was pure and done passionately. These accounts surely idealized the "good old days"; as Dr. Honjo's lawsuits make clear, however, not everything was probably so rosy. Still, it is this idealized account that we turn to in the next chapter, which illuminates what these informal and thus ill-documented practices looked like as well as how the professors who experienced the policy shift describe the past and the changes in their own practices.

THREE The Good Old Days

Trust and Ties before the Japanese Bayh-Dole Act

DR. MINAGAWA[1] WAS an old-timer in this business: he had been born in the 1930s, had worked at Harvard and Stanford before becoming a tenure-track professor in Japan, had established himself as one of the most prominent biochemists in Japan by the 1970s, and had gone on to chair many international conferences and initiatives. At the time of our interview in 2012, he was a professor emeritus at one of Japan's most prestigious national universities and the CEO of a biotechnology research tool company. In a very tangible sense, he had seen it all; he was a visible symbol of the rise of science and innovation in Japan during and after Japan's economic miracle. As a wave of biotechnology entrepreneurship had come to Japan just when he was about to retire, several Japanese firms offered him to finance a startup; he agreed to establish one that sold cutting-edge research tools.

Looking back from the present, it is hard to understand what it was like to conduct science in Japan before the 1990s. There is sporadic quantitative evidence about how Japanese academia functioned in those days, but most of these accounts are devoid of the "feel" of what it was like to be a bioscience professor in Japan. This chapter fills that gap in our knowledge by analyzing firsthand accounts from scientists who were active before the 1990s. In these accounts, scientists describe their lived experiences of practicing science and collaborating with firms. Of course, the usual caveat associated with interviews applies. Interview-based accounts are a retelling of the past from the perspective of the present and so are sure to be shaped by intervening experience—in this case, the experiences these seasoned scientists went through when Japan changed its science policy to be more commercialization-focused. Nevertheless, multiple accounts of the "good old days" enable us to imagine how it felt to be a bioscience professor between the 1960s and the 1980s. Nostalgia is part of retelling the past, but it is clear from these accounts that there was something about doing science in that era that no longer characterizes scientific practice today. Japanese university scientists worked with firms in an informal manner, downplaying the financial aspect of these professional relationships. Being a professor meant pursuing academic knowledge while serving as a scientific consultant to firms. It was assumed that relationships with firms would be mutually beneficial in the long run.

Ascetic Orientations among Japanese Bioscientists before the 1990s

During our interview, Dr. Minagawa spoke about the relationships between university scientists and firms in Japan during the 1970s and 1980s. In those days, he said, he and other scientists cared very little about intellectual property or about making money. They were all about scientific inquiry, and firms were welcome as long as they joined the community of inquiry and adhered to its principles:

> Minagawa: We'd accept firm scientists to come to our lab with the condition that they'd be working together with us for projects that our lab ought to do. I never did subcontracted [*shitauke*] research.
>
> Interviewer: Oh, I see. So what were the firm's purposes in sending these scientists, besides learning?
>
> Minagawa: Well . . . you know, at that time, Japan was experiencing a bubble economy. So they wanted to hire talented people and nurture them so the firm would be even more energetic; they had such a drive. I would say it was a good time. Thanks to that, no one thought of purchasing the fruit of university research by making professors work. [. . .] At that time, I'd go to the United States and see that so-and-so had subcontracted work, under his supervision, things like that, but that did not exist in Japan back then. It was the old style, so, we used to say that there were two kinds of research: pure and useless or impure and

> useful. Those were the two categories. And universities dealt with the useless one.

In this excerpt, Dr. Minagawa recalls focusing on "pure and useless" research. He interacted with firms, but he would only work with firm scientists if they were willing to join academic projects that he directed—not if they wanted to "subcontract" work to "purchase" its fruits. As he says, it was "a good time": the economy was reaching new heights, and Japanese firms had both motivation and pots of money. These conditions, together with the scientists' focus on "pure and useless" science, enabled the particular practices that sustained university-firm interactions during the latter half of the twentieth century in Japanese academia. There is a hint of superiority in the professor's comments about American scientists doing "subcontracted work": *shitauke* translates to "subcontracted" with the connotation of control by the funding organization. Toyota, for example, works with a lot of small *shitauke* firms that make parts. As Dr. Minagawa claimed, that kind of arrangement didn't exist in Japan's scientific community.

Academic scientists in Japan, in this narrative, had a somewhat ascetic orientation: they adopted the notion that becoming a scientist meant giving up potential financial gains. National university professors were national employees (and were still treated as such even after national universities were incorporated). Their salary scale depended on their tenure status, and the scale itself was (and still is) pretty modest. Because scientists at prestigious universities presumably could have gone into private industries—for example, the pharmaceutical industry—where salaries were significantly higher, professors considered themselves to have "given up" higher pay to pursue their academic interests. And as a professor emeritus who had become a chairperson in a public research institute explained, merit had no effect on academic salaries:

> I know the people at Genentech very well. The top-notch scientists from UCSF and other places went there—so many of them. Even now, a scientist with cutting-edge research could become president of a first-rate firm there. Due to the difference in practice—when there was no such way [of doing things] in Japan—I didn't see it as advantageous to try, and it wouldn't be advantageous to firms either. So again, this is the difference in the way of thinking among the Japanese—in Japan, professors were national employees, and even if one excelled at his work, he would still be paid according to the rules. And the professors and their families took it for granted.

This professor said he had friends in the United States who were "millionaires" but that professors in Japan didn't expect their profession to be lucrative. In addition, he said, there was little mobility between firms and academia. Because of these institutional differences, he didn't see any point in trying to do the same thing as his American counterparts by starting or joining a startup for medical applications.

The notion that professors should not be interested in financial gains is evident in these quotes. In fact, as we will see in chapters 4 and 5, Japanese professors very rarely framed their involvement with commercialization in financial terms, even if they were handsomely compensated for their entrepreneurial efforts. Before the policy changes of the 1990s and 2000s, professors seem to have exhibited a rather carefree attitude toward financial compensation. When a microbiology professor told me in 2010 that he would give a patent to a firm in exchange for a research donation of about ten thousand dollars, I was surprised and asked, "Only ten grand?" He answered, without much fuss, "In our subdiscipline, it's like that." In fact, my framing—how much a firm should pay for a patent—was misguided. After the interview, he happily gave me a couple of bottles of a dietary supplement based on his inventions—a supplement that, I knew, was pretty popular in the market and was probably yielding a significant profit for the large food and beverage firm that he collaborated with.

From the perspective of these scientists, scholars' focus on money and financial compensation in such exchanges is misguided because its roots lie in the contemporary understanding that everything can and must be translated into monetary worth. Much of the literature on university research and its applications takes for granted that the rise of university-firm interactions and patenting leads to the economization of science. Based on this assumption, this literature asks how industries buy up professors' research output and how these monetary transactions affect how universities are organized and how professors operate.

But in Japan, interactions between industry and academia didn't directly lead to the economization of science. By the 1970s, firm-university collaborations were very common, although the popularity of these activities depended on the disciplines and subdisciplines. Lewis M. Branscomb et al. show that if we judge by the number of co-authored publications, Japanese university scientists in the 1980s were collaborating with industry as frequently as their American counterparts.[2] As this finding suggests, Japanese scientists' claims about their zeal for "pure" research and their disinterest in patents are half-truths. In fact, both their science and their activities in relation to industry entailed multiple transactions and meanings beyond clear-cut financial compensation.[3] This multiplicity enabled them to draw their own boundaries and define the nature of Japanese academia for themselves. And indeed, money never seemed to override other considerations, such as "pure" science, educating students and firm scientists, and reciprocal trust between the collaborating parties.

The "new" economic sociology provides helpful guideposts here. When I puzzled over the ten-thousand-dollar patent that seemed to benefit the firm so much, I was engaging in typical academic thinking about professors' failure to price the commercial value of their research. It's more helpful, however, to focus on the relational work the professors engaged in before the policy changes introduced a new emphasis on formalized and explicitly financially compensated transactions between academia and industry. Classic anthropological works tell us that an exchange is never a purely economic transaction. In a gift economy, for example, an exchange carries clear implications about the social identities of the exchange partners—who they are in relation to, for example, family, political, and religious institutions. The thing that is exchanged has no value or meaning isolated from the nature of the exchange and the relationship between the exchange partners. Once two people enter a gift relationship, they are bound by expectations of reciprocity and gratitude. A gift exchange is a mutual expression of a relationship, such that refusing a gift, for example by immediately giving back something of equal value, would breach the two parties' tacit agreement about the nature of their relationship.[4]

We don't have to look at gift societies such as Melanesia or Native America to argue that exchanges constitute identity and morality. In fact, the new economic sociology has specifically taken on the task of proving that even in modern times, when pecuniary compensations seem to have taken over the world, people attach different meanings to different kinds of exchange. People even differentiate money—the most transparent medium of exchange—by "earmarking" it by its origin, use, and signaled relationship ties. The five-hundred-dollar check an uncle gives to the family of a newborn is expected to be used for the baby, not for buying the father a new computer; an employee may treat an unexpectedly large summer bonus as "fun money" and use it less cautiously than she uses her monthly salary, despite the fact it is all the same money.[5]

The exchange of "favors" between Japanese firms and academia before the 1990s resembled a gift exchange and involved a corresponding mutual expression of relationship.[6] In the United States, universities' formalized, contract-based relations with industries emphasized pecuniary and transactional ties. But Japanese university scientists operated under different exchange terms than their U.S. counterparts: their interactions with industries were framed less as financial transactions and more as gift giving. When Japanese university scientists described how they worked with firms, they used words such as "trust," "gentleman's agreement," and "familiarity" and described an expectation that firms would eventually return their favors as long as they continued their relationship. Their practices resembled a gift-exchange relationship not only in their pattern of delayed and trust-based exchange but also in their bundling together of various kinds of goods, including research, human resources, money, information, and tools, to the extent that university-firm interactions became an essential part of doing science for many Japanese university bioscientists.[7] For these professors, working with industry constituted a part of their professional identity—what it meant to be a professor.

All exchange relationships can generate conflict, and that includes the less financialized, less transactional relationships between Japanese scientists and firms before the 1990s. If Marcel Mauss, the classic theorist of the gift, was clear about anything, it was that gift relationships in traditional societies were imbued with violence, political struggle, status competition, and humiliation. So to say that Japanese university scientists and firms engaged in less monetized exchanges is not to say that their exchanges were devoid of conflict or contradictions. It also is not to say that Japanese university scientists had warmer, more meaningful, more fulfilling relationships with industry than were typical in the U.S. And yet, the kinds of conflicts and contradictions they faced were different from those their American counterparts faced during that time and had a broader impact on their identity, sense of honor, and social relationships.

The Institutional Contexts of Japanese Academia, 1960s–2000s

Both national and private universities have typically allotted only very modest research funds to science labs. Although grants-in-aid have become increasingly available over time, there was a dearth of research funds. Meanwhile, industries such as engineering, agricultural sciences, and medicine were undergoing rapid development and increasingly needed new research. Conditions were ripe, then, for some kind of arrangement between scientists and firms, and a lack of institutional scaffolding for industry-university interaction meant that *shogaku-kifu* donations were part and parcel of university life. This under-the-table interaction between university scientists and firms seemed to work pretty well—well enough, in fact, that the donation scheme has never truly been replaced.

As Japanese industries became more established and increasingly interested in collaborative research with university scientists, the Japanese government began to set up additional, more formal methods of collaboration between university labs and firms. During the 1980s, the Japanese government gradually crafted official, formal initiatives for university-industry cooperation. But at the same time, the government also encouraged the use of the donation scheme. Until 1983, the government officially only allowed contracted research that had a clear public benefit.[8] In 1983, through notifications from MEXT, it established a scheme for research collaborations between universities and firms with the understanding that donations would still continue to play an important role in academic life.

On December 22, 1984, MEXT issued a notification to clarify the different purposes and procedures of the donation scheme versus contracted research. The fact that these two activities were listed together in a single notification is telling: it indicates that by the 1980s, donations from firms were clearly recognized as vital sources of research funding. The notification, officially titled "Regarding the Acceptance of External Funds Such as Research Donations," formally acknowledged the role of donations even as it introduced new procedures.

> In recent years, many parties in society, such as industry, have expressed many expectations and made requests concerning the research conducted in universities for the promotion of science and technology. Because of that, we observe a trend: the yearly increase of research funds, such as research donations and contracted research expenses that national universities accept from external institutions such as private firms and public research organizations.
>
> Recognizing this situation, [. . .] the Ministry of Education plans to improve and simplify the administrative procedures for accepting external funding such as research donations.[9]

In this notification, the Ministry of Education ordered research donations to be used flexibly as an extra pot of money outside the national university budget. It also decreed that universities could use donations for purposes beyond those that were allowed for the university's official research budget—for example, to pay registration fees for international conferences, budget for organizing meetings and conferences, fund faculty and student travel, invite foreign researchers, or hire research and administrative assistants. Back then, these expenses were not easily covered by university research funds or grants-in-aid. The notification also encouraged the chairperson of the university (who is the official administrator for such donations) to delegate administrative responsibilities to the academic departments to enable swifter operation.

This notification was clearly intended to provide an administrative mechanism to enable firms to donate funds to university scientists without excessive bureaucracy and restrictions. The notification's inclusion of both research donations and contracted research is evidence that research donations were publicly understood as mechanisms for collaborative research, despite their purported status as gifts. This notification was followed by a sequence of additional notifications that cemented the Ministry of Education's intention of enabling and enhancing university-firm interactions. Then in 1987, the ministry responded to an increase in university-firm interactions by ordering national universities to establish administrative offices for collaboration. But the *shogaku-kifu* "donation" scheme continued simultaneously as an informal alternative to the more formalized approach to collaboration backed by these offices and their paper work.

After the policy changes of 1983 and 1987, scientists could have official joint research approved by their department instead of working informally through research donations, especially when there was a grant involved. But informal collaborative projects were far more ubiquitous and generally continued to be seen as legitimate. Part of the reason for the endurance of donation-based collaboration was its relative lack of restrictions. Formal collaboration contracts and collaboration through grants-in-aid both came with restrictions on who could own the patents for any resulting intellectual property, in many cases limiting ownership to the Japanese nation.[10] Because donation-funded research and research funded by professors' research allowances had no such restrictions, informal patent arrangements were possible. The failure to transition from donation-based collaboration to more formal collaboration schemes reflects scientists' and firms' shared preference for collaborations that take place as part of an ongoing relationship (instead of quid pro quo transactions) as well as scientists' belief that monetary compensation should not be a primary focus of their interactions with firms, although such compensation may benefit their labs.

Collaborative Research Based on Friendship

In chapter 1, I told the story of Dr. Koyama, who had described his experience of industry collaboration in an article for *Industry-Government-University Collaboration Journal.* According to this article, Dr. Koyama's collaborative work had begun with a simple visit to his lab from a senior director of a manufacturing company. He had collaborated with the company for years without any formal or even verbal agreement. When the collaboration eventually enabled the company to create new equipment, the senior director, who had now become the company's president, began sending Dr. Koyama research funds every year as an expression of gratitude and to fund subsequent improvements to the equipment. Dr. Koyama described his relationship with the firm as "solely based on 'trust'" and as characteristic of "Japanese culture."[11]

This kind of interaction was pretty common "back then." The relationship between the firm and the scientist was based not on written agreements but on trust—trust that each party would do good without resorting to sanctions. What does appear to be unusual about Dr. Koyama's experience is that by his account, the firm did not initially provide anything at all to support his research. By the 1980s, it was common for firms to donate money to support scientists' research. A senior professor in biomedicine described the process: "We got *shogaku-kifu* money [research donations]—this is rather personal—we were close to the chair of a firm's lab, so we'd ask, and they gave it to us. That was when there were no restrictions on how to use the donation—they'd submit the money to the nation, but the whole amount would come directly to our lab. We could hire people, buy instruments, whatever. Before, people were doing [gesturing money under the table], so they [policymakers] created this. This system is still working." As this excerpt exemplifies, *shogaku-kifu* was often given by personal friends working in firms; these friendships had often been formed through networks developed through graduate education or conferences. Although *shogaku-kifu* is formally a donation and cannot, strictly speaking, be used to fund specific projects, it was widely understood that the funding firm would receive some returns that are informally agreed upon—and the Ministry of Education's notification eventually formalized this flexible use of research donations.

These research donations were one medium of exchange between university bioscientists and their collaborating firms.[12] But they weren't the only medium of exchange, and the exchanges themselves had multiple dimensions, meant different things to the two parties, and were delayed compared with typical contract-based monetary exchanges—as is typical of a gift exchange. Japanese university scientists treated their collaborations with firms as part of their professional lives but at the same time framed them as personal—built on the relationship between the university scientist and the firm scientist with whom one was collaborating. They also regarded their collaborations as a public duty—they considered themselves to have a responsibility to disseminate their scientific and technological expertise for the public good.

Gift Exchanges between a Scientist and a Firm

When Japanese university scientists spoke with me about their interactions with industry, they often evoked the *personal* nature of their relationship with the person in the firm who managed the interaction. In some cases, it truly was a close personal connection. For example, a professor of biomedicine told me that he had met the son of the vice president of a Japanese pharmaceutical company when he was throwing out bags of garbage in the trash area of his apartment complex in Tokyo—"although his unit was the penthouse," he joked. More often, though, scientists' "personal" relationships had been fostered by professional activities—such as going to graduate school together or meeting frequently at conferences or meetings. These relationships were "personal" in the sense that many decisions about small donations were often made by middle or senior managers without further approval from the firm's upper echelons. The personal feel of the relationships was also fostered by the absence of university intervention in these collaborations and the idiosyncratic nature of the exchanges, which were guided not by a clear set of rules but by good will and the expectation of a fair return.

Sometimes, the purpose of the donation was intentionally unspecified: some firms chose to donate without asking for any research output. Many firms used research donations simply to create connections with university labs that were doing up-and-coming research. It was not uncommon for a thriving lab to receive five to ten thousand dollars a year from each of several firms for a few years consecutively. Some professors I interviewed said that at times they had received such donations from half a dozen firms every year. These small amounts of money contributed significantly to the professor's research budget but didn't really do any "work" other than maintaining the relationship and giving the firm a general sense of the cutting-edge research.

Firms often found such connections useful later on and often increased their donations when the professor provided information, services, or favors. Firms with donation-based connections to labs sometimes recruited the professor's graduate students after they finished their degree program—a common hiring practice to this day. Their donations effectively gave them a sort of right of first refusal to hire strong students. Firms also asked professors to introduce the firm's scientists to university scientists in other labs. In particular, when firms wanted to approach academic scientists abroad, they typically asked an established Japanese scientist in the same field to put in a good word for them. In addition, some firms—especially large corporations with their own research institutes or R&D divisions—enrolled their scientists in a program called "PhD by dissertation," which offered advanced research experience and the PhD without any coursework. This program enabled firm scientists to hone their expertise by working with a university professor while continuing to work full-time at their firm. Firms encouraged their scientists to participate in this program as a way of enhancing the ability and reputation of the firm's lab by nurturing its scientists as well as increasing its number of PhD-holding scientists. The program required firms to pay tuition to the university, but most firms also donated additional funds to the professor who supervised the research.

All these exchanges were explicitly multifaceted and bundled.[13] If the primary aim for the interaction was the education of the firm scientists, firms paid fees for the relevant training and the consequent transmission of scientific knowledge to the firm (because firm scientists very rarely switched firms). But firms also made small donations of research funds to the professor and sent their own scientists to the professor's lab to advance some projects. Often, firms would do both and more: pursue the education of their scientists, collaboration with the professor, and a close relationship all at the same time. In the course of describing his interactions with firms, a biochemistry professor explained how it helped his lab to have firm scientists on board:

> Another pharmaceutical thought our research was interesting, so they said they'd like to join—someone connected us—and asked to do research. They

> sent two scientists; both of them were excellent. Simply put, they could do the job a few times as fast as our students. Their purpose was to get a PhD. [. . .] They couldn't skip their work at the firm, so one person came from Friday to Sunday; the other one took longer [to finish his PhD], because he came only at night. But he would run the experiment beautifully, despite the fact that he was only there in the evenings. Compared with students, they were far more efficient. [. . .] They'd have to pay the university's fee, and they'd also given us a donation. [. . .] But at that point, the issue of money wasn't that important. It was far more important to have good people. At that time, that kind of research became popular.

This professor describes a firm sending its scientists to his lab so they could earn a PhD. The firm clearly also saw it as an opportunity to learn about the professor's research, and so they donated money to his lab in addition to paying tuition for the PhD supervision. The professor, in return, not only accepted the firm scientists as students but also appreciated them as additions to his workforce.

During interviews, Japanese professors often mentioned that firm scientists had been welcome additions as researchers and even as educators. For example, another biochemistry professor said, "To be honest, they would also teach our students how to run experiments." The nature of the exchanges and the meanings of the donations were intentionally underspecified, and the exchanges had the quality of gift exchanges: the transaction was not immediate, the relationship was meant to be long-term, and each party understood the exchange as part of an ongoing relationship.

Consulting, Donations, and Collaborations

Sometimes, firms wanted professors' scientific expertise for more specified research programs. In these cases, firms could also exchange their research donations for professors' advice. Consulting work ranged from speaking at a firm's internal conference about "whether the foam of the beer is actually beer," as one microbiologist jokingly said, to visiting a firm's lab and helping analyze experimental data. Professors had been allowed to engage in informal consulting from the beginnings of the Japanese university system, since disseminating knowledge was part of the universities' stated purpose. In the 1980s, universities began to establish more formalized procedures for consultation. Although these formal procedures did take root, especially when the firm preferred to formalize the collaboration, firms continued to use research donations to pay for consultation and collaborative research because they offered more flexible arrangements. The professor who had met the son of the vice president of a Japanese pharmaceutical company while taking out the trash ended up participating in just such an exchange: "So their junior [meaning the son of the vice president] was taking out the trash and asked me, 'Any ideas about a good drug?' So I replied, 'You are working on [a biomedicine].' And we discussed how to pick the best one from all the inhibitors they were working on, by examining the basic data. [. . .] I helped them choose one from the many candidates produced by [another company subcontracted to manufacture the medicine]. So, I don't have patents or anything, but they gave us a few million dollars." In this case, the exchange began with small talk in the trash room and developed into a consulting gig. The professor helped the firm choose the best drug candidate by looking at the data, and when the candidate he helped choose became a very successful biomedicine, the firm gave his lab a few million dollars to show gratitude. This professor exemplifies the ideal case of this consulting model: he advised the firm out of good will, and the firm donated a large sum of money to his lab when the medicine became successful.

The same dynamic characterized collaborations that included benchwork in addition to consulting. Again, although formal contracts for "collaborative research" and "sponsored research" became available in the 1980s, the donation scheme continued to be used because of the flexibility of its terms. In most cases of informal consulting and collaboration before the turn of the millennium, the professor and the firm exchanged a nondisclosure agreement but didn't specify any other conditions of the collaboration. The amount a firm donated to any one lab usually ranged from five thousand to thirty thousand dollars per year.

Because no laws required either party to fulfill any promises, scientists described these relationships as based on trust or, as some of them put it, *giri-ninjyo*, a term coined around the seventeenth century to mean "duty and humanity" that supposedly govern ethical and compassionate living. They also used the English vernacular, a gentlemen's agreement. A microbiology professor told me in 2010 that he still preferred working with firms through research donations:

Professor: In the past, you'd only write an *oboe-gaki* [letter of acknowledgment], and if something turned out [to be valuable], we'd continue, and even if not, depending on the strategic importance for them, they might say, "Let's continue half a year more." [. . .]

Interviewer: What about royalties for patents?

Professor: [Royalty] is a theoretical way of doing things, but we have the world of *giri-ninjyo*. For example, they may donate $50,000 next year, even though we have no research. They may give us favors in other ways. The biggest thing for me is that once we build a trust relationship, they'll bring research projects one after another. In our research area, we excel at exploration technology, but we can't keep producing new themes. So this human relationship, it brings various research themes; that is truly appreciated—it is priceless.

This professor's lab never claimed IP rights when an invention resulted from this collaborative research; the professor was listed as one of the inventors, but he gave the firm the right to be the patent's assignee and thus the exclusive right to use the invention or license it out. When I asked him about royalties from patents, he said that royalties were theoretical, implying that it is not a practical way of doing things. In fact, his lab usually just let collaborating firms take ownership of patents and trusted the firm to continue providing research donations. By using the term *giri-ninjyo*, he was claiming that his collaboration with the firm was governed by each party's trust that the other would behave as expected. The professor also made it clear that he found his exchanges with firms to be far more complex and meaningful than more straightforward monetary transactions were. "The biggest thing," he said, was the relationship the exchange nurtured—a relationship through which the firm would keep bringing him new and interesting projects.

Like this professor, many of the university scientists I interviewed had simply handed their IP rights to their collaborating firms. Many, though not all of them, told me that they had not become interested in patents until at least the 1990s. A professor who'd been active in biotechnology in the 1970s said, "My research was basic research, and at that time, nobody thought studying DNA would make money. I published multiple papers but never thought of patenting. . . . Thinking back, there are a few that I think we should have patented." And a prominent professor of immunology told me in 2010 that he had let the pharmaceutical firm he was collaborating with take the assignee rights to a substance patent he'd been involved with because "we didn't think about it too much": "In the university, nobody thought about it that way. Twenty years ago, we didn't even think about patenting. It costs money. First of all, universities completely lacked the mechanism to obtain or take care of patents. Now we have an intellectual property division, but nothing like that then. You'd go to the university to talk about patenting, and they'd say to do whatever you like, do it individually. So, we didn't have any reason to patent it, so [the pharmaceutical firm] said 'Let's just make sure we have a substance patent,' so they did it for us." This professor was quick to add, accurately, that many American bioscience labs didn't patent their findings in those days either. What seems to be the difference, then, is that while American academia was culturally and institutionally transformed by the birth of biotechnology and Stanford's move to acquire the first biotechnology patent, no such transformation happened in Japan.[14] Different professors and different subdisciplines had different levels of interest in patenting and different perceptions of the legitimacy of faculty patents. Still, the sentiment that patenting was not something that a scientist does was widely shared.

Even in the subdisciplines that were more engaged with industry collaboration and patenting, the assumption was that patents were the firms' responsibility. Among the disciplines of my interviewees, engineering (including early forms of bioengineering) and agricultural sciences (including microbiology) had historically been relatively friendly to patents and collaboration with firms. Professors in these disciplines recalled that when they had been students, university labs had been flooded with firm scientists. A microbiology professor said that in 1980, the lab he belonged to had an abundance of firm scientists:

> Professor: My professor even joked in the lab that if there were too many firm scientists, this would cease to be a university! The lab had so much collaboration, and a lot of the work we did led to applied research by the firm scientists. My research was not applied, but [. . .] it was cutting-edge then. So it was patented, though of course it didn't lead to an application. [. . .]
>
> Interviewer: How did you patent it back then?
>
> Professor: It was all done by firms. I conducted all the research, and my professor became the inventor, and the assignee was the firm. Then, there was no such thing as an IP division in universities, and it cost money. So including that, you'd just ride on board with the firm. OK, let's be honest, I think I actually received fifty thousand yen from that

firm (about a thousand U.S. dollars in 1980). For research that took me three years! [laughs] Though I was confident that it would not be applied successfully—it was too immature.

After the 1970s, collaborations between firms and university labs became increasingly common in the biosciences. And as these quotes suggest, the collaborating firm often obtained the patents to inventions arising from the collaboration and even inventions arising from the lab's other, noncollaborative research. The quotes summarize sentiments expressed by many of the professors I interviewed. First, they had limited interest in securing patents for several reasons: they believed that universities should prioritize basic research, they thought their research had a low probability of resulting in marketable applications, and their universities did not provide organizational support for patent applications. Second, collaborating firms were often interested in securing intellectual property and typically offered to pay for the application and management of patents. Third, as a result of these factors, collaborating firms often obtained intellectual property rights, which meant that the firm became the patent's assignee while the professor and any other researchers were listed as inventors.

Most professors were comfortable with this arrangement, because they trusted that the firm would return what was due to them based on their relationship and its associated duties and obligations. When I asked a professor of medicine whether he thought it was unfair that firms acquired the patents to his inventions, he said no:

Professor: Well, it may have been disadvantageous to the researcher, but back then there was no mechanism, so it's either the researcher pays out of pocket to become an inventor and an assignee, or just give [the patent] up to the firm, and I chose the latter. So I'm the inventor but not an assignee.

Interviewer: OK.

Professor: So the cost for the application would be paid by them, and they'd own the technology. In return, they'd increase their research donation. That was the agreement.

Interviewer: OK. You didn't feel that was unfair?

Professor: No, I didn't. I thought it was a balanced agreement, given how likely it was that the patent would be a hit and how much we needed funding for our research at the moment. I thought the chance of it leading to an innovation would be low. Inventing medicine itself is a low-probability thing. Also, it would take ten years or so, and then I would be getting royalties. If I didn't want to slow down my research, I'd rather have research funds that could be used right away. This is how it worked.

At first glance, the terms "gentlemen's agreement" and "*giri-ninjyo*" suggest that scientists were not being especially calculating when they entered into collaborative relationships with firms. But that was not quite the case: the professors I interviewed who had collaborated in exchange for research donations told me that the arrangement felt reasonable to them. Of course, as in any economically entangled relationship, power relations between the two parties were evident.[15] Professors did not have the money or institutional resources necessary to protect their intellectual property, nor did they get any direct reward from the patents associated with their inventions (except in the unlikely case that the patent became essential to a firm's product). In this situation, many Japanese bioscientists felt that it was reasonable to give their IP rights to firms that were willing to apply for them and respond with donations to the professor's lab. Many of them thought that this might be the best deal they could get anyway.

The policy intervention of the 1990s was not motivated by university scientists' protests. But some university scientists I interviewed did criticize the earlier gift-exchange system of collaboration. Some had been less than happy to give their IP rights to firms and said that the firms "used to buy us cheap" or even exploit professors. A biomedical professor from one of Japan's top universities expressed his sense of this exploitation:

Professor: The firm scientists I met twenty years ago are now all in top management. I've known them for twenty years, which means they have squeezed me for twenty years to use my ideas. This is highly unachievable. They don't talk about what they are doing in research; they only take from me. To continue that for twenty years is astonishing.

Interviewer: That's quite right. How is that possible?

Professor: I have no idea. [laughs] When they come visit me, they aren't here to hang out. [. . .] So the old way of doing things was really "one box of sweets" [that the firm brings, to take professors' knowledge for

> free]. Having a system so scientists could have returns would be good, I suppose, although if we claim too many rights, that would cause trouble too.

This professor expressed negative feelings about the old approach to university-firm collaboration. He clearly saw his relationship with firms as unequal; in his eyes, the firm took and the professor gave. He used expressions such as "squeezed" or "box of sweets" to emphasize how exploitative the firms were. At the same time, he said that if universities and university scientists claimed too many rights, it would be detrimental to commercialization.

Perhaps because of his ambivalent stance, this professor chose not to pursue alternative methods of exchange with firms. And such alternatives were available, though as with many traditional forms of exchange, the alternatives were few, and the default was taken for granted. If a professor was dissatisfied with the tacit terms of the exchange, he couldn't easily negotiate the amount of the donation, nor could he easily find another firm that would donate larger amounts. However, some professors—especially prominent professors—did devise other ways of working with industry.

When the Gift Exchange Goes Awry: The Negative Side of the "Good Old Days"

Was the old donation-based system of collaboration an unreasonable, unfair practice? The answer ultimately depends on what the professors and firms thought the nature of their relationships was, what the public thought a professor ought to do (since most Japanese universities were and still are heavily subsidized by the government), and the actual rates of return—that is, the likelihood that the patent would become profitable. Robert Kneller and Sachiko Shudo argue that the old Japanese system created many unused patents, because the intellectual property had only one opportunity to become useful.[16] If the collaborating firm that took the patent did not utilize it, there was practically no other way for the patent to be utilized unless the firm was willing to license the patent to another—possibly competing—firm. By contrast, if the university owned the intellectual property rights, it could license the patent to many firms, not just the firm that had collaborated with the professor.

Dr. Honjo, whose research led to the commercialization of the PD-1 inhibitor known as Nivolumab, would agree with this criticism. As chapter 1 showed, Dr. Honjo was frustrated with what happened in his collaboration with Ono Pharmaceutical. Ono is an assignee—an owner—of the intellectual property for the cancer immunotherapy precisely because Dr. Honjo and Ono Pharmaceutical had worked with each other for a long time, beginning long before the new formal structure of university research commercialization was introduced. Dr. Honjo expressed frustration that Ono had not responded to the medication's success by donating to Kyoto University and in his comments alluded to the unfairness of working with a firm through informal donations.

Before the policy change of the 1990s, professors who did not like the then-default arrangement had a few courses of action available to them. First, some (but not many) professors patented their inventions out of their own pockets. This meant that they had to deal with the paper work by themselves, without any help. But because they had no resources to market their inventions, this course of action usually ended with them abandoning the patent after a few years. "It was tuition for learning about intellectual property," one professor commented after telling me that he had patented his invention only to let the patent lapse after a couple of years. Second—and this was probably the more common path—professors could negotiate for some kind of formal IP rights, such as being one of the patent's assignees, while asking the firm to pay for and administer the patenting process. In the co-assignee arrangement, the professor and the firm would be joint assignees of the patent, whereas in the default practice, the professor and all parties that had been involved in the invention would be listed as inventors. Table 5 shows the difference between the arrangements.

This arrangement of becoming a co-assignee of a patent was more common in the medical field than in the other sciences. A professor whose inventions had become fundamental to several biomedicines recalled how he and the collaborating firm had arranged IP rights in the 1980s and 1990s:

> Professor: Firms, especially in engineering, have thousands of patents, so it makes sense to let patents sit, unless someone infringes on the patent. But for medicine, for [the pharmaceutical company he collaborated with], one patent is everything. That single patent counts. And you'd need to be an assignee. University professors didn't know, I didn't know, at first. So even if you were collaborating, without patents it was meaningless. So patent every single thing if possible, and when doing so, you have to be an assignee; if you're an inventor, it's all the firm's—they can start or stop [commercialization]. The most crucial thing is,

TABLE 5: Possible arrangements of IP rights before the policy change

	Assignee (an entity that has property rights to the patent)	Inventor (the person or persons who contributed to a patentable invention)	Exchange	Notes
Firm-owned patent	A firm that collaborated with the inventing professor or that agreed to take ownership of the patent	The professor and the professor's teammates, and if the firm contributed to the invention, the firm's contributing scientists	Long-term favors, such as recurrent research donations, consulting, collaboration, and student recruitment. Larger returns to the professor's lab if the patent contributes to a product	Most common before the policy change
Firm-and-inventor-owned patent	A firm that collaborated with the inventing professor or that agreed to take ownership and the professor as co-assignees	The professor and the professor's teammates, and if the firm contributed to the invention, the firm's contributing scientists	Same as a firm-owned patent, except the return may be pre-arranged and legally enforceable; the professor reserves the right to exercise the rights to the patent	Less common, but some professors, especially many in the field of medicine, chose this practice. Required the professor to negotiate terms with the firm
Inventor-owned patent	The inventor (the professor)	The professor and the professor's teammates	None; the inventor-professor pays and manages the IP rights	Less common, used when the professor invented without a firm's help and did not find a firm to take ownership of the patent. The professor paid for the IP rights out of his personal money.

if [the firm] said they'd continue working on [the medicine candidate] but changed their mind, nobody on our side would know.

Interviewer: Yeah.

Professor: So it can be half-and-half in the assignee position. If it's half the firm, half me, then you can say that if the firm isn't doing anything, I'll work on it. Holding that right is important.

Interviewer: Right. But how did you know about it? Many people don't know these things.

Professor: Well, from somewhere. "Make it a co-patent"—so I was told. Otherwise it wouldn't work. But we would write a memorandum that said that if the patent became meaningful, they could use it.

This professor realized he could negotiate with the collaborating firm—a pharmaceutical firm—to be one of the co-assignees of a patent that later became highly marketable. The advice to become a co-assignee had come through word of mouth. For many established professors, this kind of advice came from their own or their colleagues' international experiences (most often in the U.S.), which will be the topic of the next chapter. Once a professor realized he *could* negotiate, he was able to become a co-assignee of the intellectual property.

Multiple parties could own the right to exercise a patent. When both the firm and the professor were assignees, the firm had the right to use the intellectual property free of charge, while the professor reserved the right to exercise the patent if he so chose. This right did not put the professor in a particularly strong position, because professors almost never exercised their IP rights themselves and could not legally license patents without the agreement of all assignees.[17] But this arrangement gave the professor at least some control and was usually accompanied by an agreement about compensation for nonuse that was more formalized than the informal system of patiently waiting for research donations. In some rare cases, professors even managed to obtain intellectual property rights in their name only while making the collaborating firm pay for the patent's administrative fees. Typically, this arrangement involved an agreement that only the firm would have the right to license the patent.

Clearly, then, depending on the patent's perceived value, a professor could break out of the old trust-based, potentially exploitative method of university research commercialization and negotiate better terms for himself. What was at stake in these cases, though, was his identity: because the ethos was that "men of the university" pursued science for its own sake without thought of financial reward, professors who negotiated the terms of patents risked being seen as mercenary. To mitigate the potential impression that they were primarily looking for compensation, professors who negotiated the terms of collaboration had to go the extra mile to explain why they were interested in commercializing their inventions. These explanations never addressed the fairness of pecuniary compensation; instead, they emphasized the need to control the process of commercialization.

Before the policy interventions of the 1990s, Japanese professors worked in an academic environment that valued fundamental research, disinterestedness in monetary compensation, and research and inventions that had a communal orientation. If they had a patentable invention, they could be interested in creating a product from it but should not linger too much on how much they would be paid for it or who was commercializing it. The research donation scheme, the gentlemen's agreement, and firms' ownership of professors' inventions served as taken-for-granted, go-to ways to collaborate with firms and obtain intellectual property rights—although that meant that most of the time firms would own those rights. In exchange, collaborating firms donated modest but usually steady levels of research funding that benefited the lab's day-to-day research operations. Through these methods, the old trust-based, gift-exchange-like relationship managed to sustain itself. The policy changes described in the introduction were explicit interventions into these personal, more gift-exchange-like practices. But these interventions did not come from professors; they were the results of a decisively top-down, America-inspired policy decision. The next chapter traces how these policy changes shaped new practices of engagement between university professors, firms, and university administrators and how the old practices, rules, and relationships shaped these new practices.

FOUR Ambiguity and Loose Coupling

The Gravitational Pull of the Old Practices

> It [developing biomedicine] takes a long, long time. So human relationships matter. Even in Japan, firms these days want to have nondisclosure agreements prior to talking. On the bench level, we make no progress that way. If you trust someone in a human relationship, you can have any discussion. That would be a more comfortable way of working for me. These days, firms are keen on nondisclosure agreements, and I'm not used to working with firms like that. [. . .] In Japan, pharmaceuticals can produce amazing drugs by working with university talent in a good way. In the U.S. or Europe, for better or worse, it's all about contracts and licensing out—all about money. But if you care about human relationships, [pause] life science takes quite a long time, so the Japanese culture of working together for a long time can be another path. There may be a Japanese university-firm collaboration. (Interview with a professor in biochemistry)

THE NEW POLICY WAS intended to improve the commercialization process. By formalizing and involving universities in the commercialization process, the policy changes created a pathway for university scientists to commercialize their research without having to rely on firms and eliminated ambiguity in relation to responsibility and ownership. As the quote here shows, however, many professors didn't embrace these changes. Some accepted them as the "new normal" at least on paper but privately lamented the loss of mutual trust associated with the long-established gift-exchange-like practices. None voiced overt objections to the changes, but many, like the professor quoted previously, talked in a more nuanced way about their sense of the old procedures' superiority. More important than their verbal reactions, though, are their behavioral reactions. University scientists did not simply switch practices and begin to work in a completely new way. They had their doubts about the new rules, and they'd had positive experiences with the old informal practice—a practice that was still available to them. So instead of simply transitioning to the new procedures, they created a set of new practices that were only loosely coupled with the new formal rules.

What does loose coupling mean, and how is it accomplished in our case? The statistics provided in chapter 2 offer some insight into the extent of loose coupling: collaborations using research donations and firm-owned patents have remained common even after the introduction of the Japanese Bayh-Dole Act, and the prevalence of university-firm co-patents, so rare in the U.S., additionally suggests loose coupling. This chapter explores how this loose coupling happened. New institutional theory points to local actors' need to simultaneously conform to norms and get their work done. When these two requirements are not compatible, organizational actors often decouple or loosely couple the rules that they ostensibly adhere to and their actual work practices. Thus, loose coupling is a vital link between an organization's adoption of formal rules and the organization's day-to-day activities—and how the formal rules get institutionalized in a new place.[1]

This chapter describes how professors navigated the implementation of the new rules for commercializing their inventions. It shows that their navigation of this new system was strongly shaped by the older informal, gift-exchange-like relationships with firms that the last chapter described. The scientists found two ways to loosely couple old practices and new regulations: they followed the formal rules in part while still maintaining some of their previous practices, and they reappropriated the new rules to partially replicate the old practices. In the first case, they achieved loose coupling by juxtaposing the old practices and the new rules. In the second, they worked within the new rules but used them strategically to make the resulting practices resemble the old ones. Because most professors used both tactics, the new practice of university research commercialization ended up strongly resembling the "good old days" described in the last chapter.

What enabled the loose coupling was not just individual professors' efforts to appear legitimate while finding what would work under the new regime; the details of the policy's implementation significantly contributed to the decoupling. As Eleanor Westney showed brilliantly in her examination of the Meiji government's effort to quickly create the modern nation-state structure, the selective and patchworked imitation of Western institutions was itself an innovation—an innovation that made use of locally available institutional elements and conventional understandings of social roles among individuals, organizations, and institutions.[2] For both the Meiji government and university research commercialization, innovation and the resilience of the old system worked as two sides of the same coin. The result of the patchworked imitation was original and innovative: in some ways it strongly resembled the old structure, and in other ways it resembled the borrowed structure. In the case of research commercialization, this combination enabled firms and professors to maintain the gift-exchange-like relationships they had cultivated in the "good old days," even after the new policy changes took full effect.

One example is from the head of one of the best microbiology labs in Japan. Ever since Imperial Universities were established in the late nineteenth century, Japanese university microbiologists have been collaborating with industry to produce agricultural products such as sake, soy sauce, and miso. It is not by coincidence the first university-industry collaboration on record in 1909 was the establishment of Ajinomoto, which uses fermentation processes to produce monosodium glutamate. This particular professor's lab has been working informally with Japanese food, agricultural, and pharmaceutical firms for more than a hundred years. In his words, university-industry collaboration became "harder to do" after the policy change. He describes his predicament:

> In this kind of research, the patents around it have already been acquired by firms. So [the university] having one patent out of a whole portfolio wouldn't make sense. Having said that, in some cases, I like to give back to the university. So my theory is that we don't have to treat ten patents all the same. Some I arrange to be one-to-one patents right to the university [the co-ownership of the rights among the university and the firm], some 80% to the university. But the university has only one standard form, so their response used to be so slow. Now they have meetings twice a month though . . . I hated the delay and just patented in the old way [giving the rights to the firm]—but the university is getting cleverer these days.

This microbiologist's case exemplifies the frustration and distance many university scientists felt in relationship to the university, which he describes as intruding into his long-standing relationships with collaborating firms. From this point of view, university ownership of IPs did not help commercialization, both because of the university's slow pace and because it's detrimental to commercialization for a university to own a fraction of the patent portfolio associated with any given product. When a product involves multiple patents, he implies, firm ownership of the entire set of patents facilitates a smooth transition from invention to commercialization, so he continued to give his patents to collaborating firms. This was how his lab had always operated, he said: "The lab directors were never interested in money. They probably never kept a patent for themselves even once. They gave them all away."

But as time went on, this professor began to feel pressured to abide by the new rules, so he started to incorporate them into his practice partially. For some patents, he arranged for the university to be a joint owner with the firm. For others, he stuck to the informal practices and asked the firm to work with him through research donations. "We ask firms to give us research donations if possible. Some understand the need and make internal decisions in favor of it, but in some cases, their legal division says no," he said in 2010. This professor doesn't care about retaining IP rights or being directly compensated for them; he prefers the flexibility of research donations.

The university scientists who were most affected by the policy change were those whose labs had collaborated with firms informally many times under many successive lab heads. This professor managed to keep using the old donations scheme and keep giving IP rights to firms, but the new structure of formal collaboration and co-patenting was starting to take root in his lab regardless of his preference.

Japanese university bioscientists' responses to the new rules were not uniform; they ranged from trying to keep using the old practices as much as possible to only using the new formalized ones. Most fell in the middle of this continuum. When scientists had already established a relationship with a firm using the informal, gift-exchange-like practices, they tried to keep using those practices until compliance with the new rules became an issue—either because they began to feel uneasy about continuing to use practices that were losing legitimacy or because the university began to monitor the process and institute repercussions.

When professors did adopt the new formal structure of commercialization, they did not use it in every case. They typically used both new and old practices and by doing so created a self-monitored "balance" between being legitimate and being old-school. For the professor quoted previously, giving the university some IP rights was good enough to satisfy his feelings about legitimacy. When I asked whether there could be any repercussions for cases where he chose to give patents to firms, he replied casually, "Well . . . there's no punitive clause, and the university doesn't have money. So it all comes down to securing that [money to apply for and maintain patents]." In his mind, because he now arranges for some of his patents to be university owned, he has paid his due to the university, so it's acceptable for him to give patents to collaborating firms in other cases.

A professor of biochemical engineering from one of Japan's best national universities described a similar strategy of ignoring the new guidelines. He used both collaborative research contracts and donations, and his words again reveal a sense of distance from his university's TLO (technology licensing office). He told me that he typically remained an inventor, not a patent owner, unless the patent seemed potentially commercializable, and he showed little worry about bypassing the university:

> The university, since it got corporation status, and they are out of cash—they all created TLOs and all, [but] they don't do any good work. For firms, they're just a hindrance. [. . .] Recently, we gave all [the patents] to firms. . . . We go through them [TLO], but we'd write the form in such a way that the university wouldn't take interest [in securing IP rights], and then they'd reply to us saying, "Do whatever you want." A guy from the IP division would come, and we'd suggest that it wouldn't have much commercial potential, and then they'd reject it. They are inclined to say "no" to start with, because they filed too many patents for a while and they're incurring cost for it, so patents are being thrown out.

This scientist's approach to the new procedures bordered on deception: he depicted going through his university's patent-appraisal process but describing his inventions to make them sound unappealing. Once the university refused to patent the invention, he was free to do whatever he wanted with it—including giving it to the firm he was working with. Both he and the professor quoted previously took the firms' perspective: the first by saying it didn't make sense for the university to own a part of the set of a product's IP rights, and the second by suggesting that the invention wouldn't be worthwhile for the university to pursue. Both professors referred to their university as "the university," and both described their university as intervening in their relationships with collaborating firms—and they clearly evaluate that interference negatively. The trust relationship between the professors and the firms was dyadic, and the policy change not only added an unwelcome third entity to the relationship but also forced the formalization of an informal relationship.

The persistence of the old informal practices raises the question of legitimacy—that is, the professors did not feel too uncomfortable to use informal practices. Professors who were used to the informal practices openly complained that the university and the new formalized rules were cumbersome, time-consuming, and meritless. In their view, the new rules were only partially legitimate because these rules did not help the scientists or firms to commercialize their invention. Their existing practices, on the other hand, were seen as partially legitimate because they were useful to them in their interactions with industry. These professors were not active resisters of the new structure; they acknowledged that the new rules were here to stay. Instead of actively resisting the new structure, they partially incorporated it into their existing collaboration practices. The result of this partial incorporation was a bricolage of practices that included the old ones and new ones based on compromised, loosely coupled versions of the new rules.

Loose Coupling through Modification: The Birth of University-Firm Co-Patenting

Even when scientists seemed to follow the formal rules perfectly, they could, and did, undermine them by following them in a partial manner or negotiating within them to such an extent that the original intention behind them remained unfulfilled. Loose coupling within the rules was particularly useful when the professors thought it would be difficult to use the informal, donation-based practices, either because university announcements made them feel pressured to adhere to the new rules or because collaborating firms that were aware of the new rules grew wary of the old practices. The degree of such pressure varied from university to university and from firm to firm, but it became increasingly difficult during the 2000s and 2010s to completely neglect the rules. This social control was not formal—there was no punitive clause or lawsuit—but informal: the rules were widely broadcasted both within and outside the university, and public discourse on universities came to embrace commercialization and entrepreneurship through promotion of the new policy. To "solve" the tension between what the formal rules allowed and what they were accustomed to and thought was sensible, bioscience professors then negotiated within the rules to create arrangements that both honored what they thought firms should be entitled to and demonstrated compliance with the new procedures.

One way scientists negotiated within the rules was by following the formal procedures only when it became clear that their collaborations would yield potentially valuable inventions, so that in those cases, the university would be officially involved in the research and IP allocations. As a professor of bioengineering at a national university stated, "For a feasibility study, if it is collaborative research, for the first six months we'd run with only a nondisclosure agreement with donations, with minimum paper work. And we'd do collaborative research if something was to be found." In an effort to maintain some flexibility of research timing, topic, and budget, this professor began his collaborations using the old practices—receiving donations and starting research with minimal paper work. Only when the collaboration took off would he and the firm contact the Office of Research and construct formal contracts. In this way, he was able to work with firms in a free-flowing manner, receive donations, and avoid the delays caused by negotiation and paper work between the scientist, firm, and Office of Research. When they thought the collaboration would be substantial and yield valuable IPs, they then went through the new formal route of having contracts for the sake of legality. This approach was commonly mentioned as a legitimate practice that follows the new rules, although on paper, all and any collaboration should have a collaboration contract through the university prior to any research activity.

Another form of loosely coupled rules and practice is the university-firm co-patent. Co-patents are tightly coupled in the sense that they are perfectly within the terms of the new regulations, but they are loosely coupled with the regulations' original purpose. The Japanese Bayh-Dole Act and university regulations were passed so that the IP rights emerging from university research would be owned by the university. In Japan, however, the default arrangement became co-patenting. In fact, the university scientist would often negotiate *for the firm* so that the firm would get as many IP rights as possible, and university TLOs and administrations were lenient about giving the firm half or more of them. As a result, university-firm collaborations often yielded what are called (interchangeably) joint patents, co-patents, or co-assigned or co-owned patents. Co-patents have more than one assignee—that is, more than one entity owns the intellectual property. In this context, the university where the professor works and the firm the professor collaborated with are both assignees. A bioengineering professor described the process through which TLOs decide whether and how to apply for a patent arising from faculty research:

> Professor: So ideally the TLO has the ability to judge immediately which inventions will be successful, but in reality, they have to do research.

And what that research means is that there are three options. The first step is the faculty member submits an A4 one-page report with necessary details to ask the TLO to obtain patents. Then the first possibility is that no one seems interested; the university will not assume the rights; do as you wish. Option 2 is that the TLO finds that there are interested firms, so let's patent. Option 3 is that when I submit the report, I'd say that this was in collaboration with firm A, or firm A would like to obtain the patent with the university. Then that means that firm A will bear the cost of patent application. So that would either mean co-patenting [with the firm and the university as assignees] or I'd remain as an inventor and the firm would apply for a patent. So in this option, the TLO won't have to bear any cost, so if we bring it to the university, the university would assume [the right to obtain the IP].

Interviewer: So in option 3, the firm will have half the patent rights?

Professor: Yes, or all of them.

Interviewer: And these are the cases where there were already collaborative research contracts?

Professor: Of course, if you have collaboration research contracts, only option 3 is available. Even when you don't have them, at some point—for example, at a conference I'd ask a firm if they are interested, let's say. And if the firm said yes, I would put in the report that firm A would be interested and would bear all the costs of patents.

Applying for a co-patent with the firm became one of the most commonly used practices in many TLOs to obtain IP rights. According to this professor, all inventions arising from university-firm collaborations are either co-patented, with the firm and the university as co-assignees, or owned by the firm. In fact, even when the invention is based on research that didn't involve private companies, the professor can still ask firms if they are interested in obtaining the associated IP rights. In these cases, the firm covers the entire cost of obtaining and maintaining IP rights and is typically listed as a co-assignee with the university.

Two points must be made here about the practice and legal ramifications of co-patenting. As briefly explained in chapter 2, university-firm co-patents are distinctively non-American. In both the Bayh-Dole Act and the Japanese Bayh-Dole Act, universities *may* own the intellectual property arising from university research (to be more specific, IP arising from nationally or federally sponsored money). In other words, neither the U.S. nor the Japanese version of the act *forced* universities to own the rights to university inventions; they simply enabled such an arrangement and left the details to be negotiated. In the United States, IPs arising from collaborations with firms are typically owned by the university only, while the firm has the right of first refusal to exclusively license the IP. U.S. universities do not allow firms to obtain IP rights to inventions arising from faculty research, and a legal agreement is a prerequisite of all university-firm collaborations.

In Japan, by contrast, firms can negotiate to become a co-assignee of the patent, and they very often do.[3] Co-patenting is legally possible in both countries, but the rights of the co-assignees differ. In both countries, a co-assignee is entitled to exercise the intellectual property rights at no charge, as they legally own the patent. But differences in Japanese and U.S. patent laws make this arrangement even more advantageous to the firm in Japan. In the U.S., a patent assignee may license their portion of the IP rights without other assignees' consent, but in Japan, licensing out the IP rights is possible only when all assignees agree to it. As a result, a firm that is a co-assignee essentially has an exclusive and royalty-free right to exercise the patent, though it may be obligated to pay the university "compensation for nonuse." The university, meanwhile, only gets the honor of having the patent—they cannot license it out without the consent of the firm.

The result of this practice, then, is very similar to simply giving IP rights to firms, because the firm funds the acquisition and maintenance of the IP rights and has the full right to exercise the patent free of charge unless prior agreements have been made. Because the university will not exercise the IP rights and cannot license them out without the firm's agreement, the university typically asks for what is called "compensation for nonuse"—which is payment from the firm to the university for their exclusive use of the patent while the university is unable to benefit from it. The only difference between this system and the informal system is that instead of receiving a research donation to the professor's lab, the university receives a (in theory) prenegotiated sum as compensation for nonuse of the patent.

Even when professors abide by the new rules, then, the exchange between the university and the firm may look very much like it would have before the policy change. To highlight this, let's take a look at a professor who only uses the new collaboration scheme. An assistant professor of nano-scale biotechnology at a top national university told me how he felt about formal collaborations. "By the time I became a full-fledged professor, the situation in the university had already changed with the Japanese Bayh-Dole and all," he said. "So I wasn't really involved in research with donations, except I remember that when I was a [graduate] student." Without a basis for comparison, he feels that the collaborative research contracts work just fine for him, and he regularly works with his university's TLO to arrange the agreements. When I asked how he dealt with the TLO regarding compensation for nonuse, he explained,

> We don't decide the share of IP rights at the time of contract. If one starts to lay out everything in the contract, nothing gets done—so these will be discussed later. Then the clause about the compensation for nonuse is what the [university name], or, to be exact, the TLO puts together. And that's how they are fed. It may sound banal, if you put it straight like this. Because, if the technology needs more long-term R&D, the firm will not produce a product—there is no result in the first couple of years. So asking them to pay compensation for nonuse is not right, right? Essentially, these payments should go through when there is a profit. So that's why I don't think the compensation for nonuse is right, so I try to get it out of the contract.

Like the other professors quoted in this chapter, this professor takes the firm's perspective: the firm, he argues, should not be required to pay the university until the profit has been made. The university administrator, by contrast, asks for compensation to make their own office sustainable. When scientists use the new procedures, they often negotiate for the firm and exhibit a willingness to forgo the potential benefits of a contract that is favorable for the university. This professor explained how he had strategically maneuvered for one collaborating firm so that it would have as much leverage for IP rights as possible:

> In [firm name]'s case, the TLO compromised. Because when it [contract negotiation] takes so long, it becomes a game of who caves in first. The most important part is the basic collaborative research contract. The firm says, "We are ready; when shall we start?" And if that contract is delayed, the budget won't kick in, so the scientists are in trouble. So we can push the office: "What are you doing to us?" Then, since the university is weak in negotiations, they'd allow us to have the contract on budget before detailing out everything else—they'd want the money to come through within the fiscal year so that the TLO is funded as well. . . . And then they'd have to compromise later, because things are moving. And once we create the precedent—the world revolves around precedents. [laugh]

Although this nano-scale biotechnology professor did abide by the new rules, he negotiated within them, and the resulting form of collaboration resembled the old practices. In fact, compensation for nonuse had been a point of contestation since the new policy was implemented because firms were accustomed to owning patents and preferred to avoid paying fees to the university. The professor in this case helped meet this firm's demands. He managed to drop the compensation clause by beginning the collaboration without developing the details of IP rights allocations so that the university would later be forced to compromise. Although he only started to commercialize his research after the new policy was instituted and used the new formal procedures, the outcome of the agreement between the university and the firm had more in common with the old practice in Japan of giving IP rights to firms than it did with the typical practice in the U.S., where universities have typically had sole ownership and management of university-born intellectual property.

We have seen two ways that professors decoupled the rules from actual practices without completely breaching the new rules. First, professors used the new rules on some occasions but not others. Juxtaposing the old way of collaborating with firms with the new, more formalized methods helped professors manage their trust relationships with firms while keeping their actions at least partially legitimate. Second, when professors worked within the new rules, they accommodated and responded to firms' interest in securing IP rights. What is striking in these cases is that professors seemed to prioritize the firm's needs more than the university's or even their own needs. This was true even when they abided by the new rules and, in the last professor's case, despite never having participated in the old donation-based system. The model of trust-based, dyadic relationships with firms was transplanted into the new system. As a result, Japanese scientists did not particularly want what would seem to be advantageous contracts for the university; they even treated their TLOs as invading their collaborations with firms. They preferred to negotiate an arrangement that would be flexible and also enable them to "give favors" to firms. By and large, then, the new rules took root in professors' practices by loose coupling, without upending the old practices of using donations and giving the collaborating firm their IP rights.

The original intention behind the policy change was to make commercialization more straightforward and attractive to university scientists by eliminating informal, unclear conditions that often favored firms over scientists. In that context, these scientists' actions may seem puzzling. They felt that they benefited from receiving recurrent donations, so does that mean they were simply pursuing their interests of receiving recurrent donations? There are two interrelated answers to this question. The first is yes: scientists may have chosen to continue the old practices or to hand off their IP rights at least partially due to ambiguity in how beneficial the formal path would have been. Even in the U.S., where university patenting is so celebrated, most university-based inventions just incur costs.[4] Given the risk, working through a trust-based gift relationship could be the most rational course of action.

But the second answer to that question is no: the scientists I spoke with *never* framed their decisions as motivated by self-interest. In fact, as we saw in the last chapter, scientists seldom framed their collaboration practices as acts of self-interest, either financially or more broadly. Instead, they emphasized the "gentlemen's agreement" and "trust" in their collaborative relationships with firms. In this frame of understanding, it didn't "make sense" for universities to pursue formal IP rights, and the university administration was a self-interested third party that the professor was not a part of. In other words, university scientists' decision to create juxtaposed, loosely coupled practices was based on a relational calculation as much as a rational one, and according to this relational logic, it made more sense to offer firms favors, typically in the form of IP rights to inventions.

Perhaps what's surprising about this loose coupling is how smooth and taken-for-granted it became to not only the university scientists but also the policymakers and university administrators—including TLOs, which did not advocate for the idea of universities as sole owners of IP rights. To understand the widespread taken-for-grantedness of this loose coupling, it helps to look more closely at how the new policies and procedures were established. This closer look shows that loose coupling was taking place not just on the individual level of university scientists but also at the levels of policy and administration. At each level, we see compromises and contingencies leading to more compromises and contingencies as the policy became its own set of administrative rules, organizational practices, and individual practices.

The Loose Coupling of Policies Generated the Loose Coupling of Practices

Like a lot of practices that become quickly taken for granted, we do not know the exact moment when the practice of pursuing university-firm co-patents became institutionalized. The Japanese Bayh-Dole Act surely enabled such an arrangement, but for that matter, the U.S. Bayh-Dole Act also enables such an arrangement, as would any legal framework that enables universities to own IP rights. This arrangement seems to have been institutionalized while the new formalized structure of university research commercialization was unfolding. In other words, there was no clear mandate or promotion from the government or the university system to adopt co-patenting as the standard practice. Instead, the expense of claiming and maintaining IP rights, the new rules' neglect of the old practices, and the "pull" of the old practice of giving the IP rights to firms all contributed to the institutionalization of university-firm co-patenting.

On the policy side, the Japanese Bayh-Dole Act was implemented first as a part of a general law designed to revitalize industry. The act meant that the university could now own intellectual property arising from university research, and its goal was to promote the practical application of nationally funded research. On paper, inventions arising from university professors' research were national inventions and officially belonged to the national funding agencies.[5] Practically, two issues that had to be resolved tilted the policy discussions and made the practice of co-patenting favorable: the cost of patent application and maintenance, and the fact that universities needed firms to make use of the patents they owned.

The Japanese Bayh-Dole Act, the TLO Act, and later the National University Corporation Act were all intended to promote research commercialization by enabling universities to own and profit from their IPs. But how to make such an arrangement possible in the context of national universities—whose budgets come directly from MEXT—quickly became a point of discussion during the negotiations prior to the establishment of the acts. The TLO Act helped by subsidizing the cost of establishing government-approved TLOs and operating them for five years. But it was unclear how the TLOs would continue to operate after this aid expired, especially since national universities were expected to gain corporation status and, due to neoliberal "reforms" that were instituted, to face budget cuts of nearly 1% of operation costs per year. In the past, because professors had routinely given their IP rights to firms, universities had not needed to budget for such expenses.

In addition to the financial dilemma, there was another consideration: the principle that university administrations should facilitate commercialization. The new policies were established based on the assumption that professors' inventions were not patented—or if they were patented, that the patents were largely left unused. This assumption was based on the lack of a formal, on-paper procedure for patenting faculty inventions. But in reality, as previous chapters have shown, faculty research was frequently commercialized through informal university-firm collaborations. The real problem with the old practices was not that firms never used faculty inventions, but that because most faculty inventions were owned by one single firm, the invention would never come to market if that firm did not pursue commercialization.[6]

Unlike policymakers, university administrators knew about the informal practices that had long guided the commercialization of faculty research. And because they knew that the policy changes were intended to promote commercialization, not to enrich universities, university administrators did not pursue full ownership of faculty inventions. Instead, co-patenting became popular as an approach that seemed to take into account the previous practices, the financial constraints, and the respective needs of university administrations, university scientists, and firms. By 2004, when national universities were turning to the task of gaining corporation status, the General Science and Technology Meeting expected that co-patenting would be one of the most popular arrangements.[7] Although the policy clearly stated that a variety of options should be available when a university and a firm collaborate, the "standard" form that MEXT distributed to national universities in 2002 stipulated that IPs arising from collaboration must be co-owned.[8] Thus, from the very beginning of the process of importing the Bayh-Dole Act and related measures from the U.S., these rules were modified to be executable in the landscape of Japanese academia.

The Response of University Administrations

University administrators and TLOs, for their part, are very lenient about giving firms co-assigneeship of IPs and give a tacit nod to the fact that firms still own many university-born IPs. As the professors in the previous section mentioned, there is no clear punitive clause for professors if they do not report their inventions. But the larger issue is that the new policy doesn't ban the old informal practices, because they've been considered legitimate for decades and because of their informal nature. In fact, if the goal is to enhance university-firm collaboration, as the policy states, it is unclear why anyone would completely ban the old practices. Even the university administrators I interviewed were typically ambiguous when I asked about going after the professors to enforce the new rules. For example, this was how the TLO general manager at one of Japan's best national universities replied when I asked whether all inventions were reported to the university:

> It's intricate whether it's enforceable. As a general rule, after the National University Corporation Act, universities set up employee invention rules, so the faculty is obliged to file their inventions with the university. And then, the university's invention evaluation committee discusses whether the university should apply for the patent. If that's the decision, the university will apply; that is the scheme. Right. So if you ask me, it's enforceable to the extent that there is the rule about employee inventions, but, you see, there is no punitive clause. So it depends on the awareness of the faculty, yes.

Similarly, I asked the head of the university-industry collaboration division of a top national university about the possibility of a professor continuing to use the old practices and not patenting through the university; he replied, "Well, that won't be visible to us. How are we to recognize the invisible . . . if that happens? I wouldn't say it doesn't. As a university, we are basically not concerned about that," adding that his university rejected about a third of faculty IP applications anyway. Patents that do not go through the university's formal processes can be ignored, he implied, since the university might very well have declined to help with the patenting process anyway. In essence, then, the university allows for the informal practice of giving IPs to firms. Even donations, he said, would be encouraged:

> The [industry collaboration division of the university] has been established for five or six years. In departments like medicine, pharmacology, or engineering, they have long been engaged with university-industry collaboration. We are a newcomer. Of course, no one talked to us in the beginning. . . . We have analyzed what are the collaborative research projects our division can encourage. [. . .] Prior collaborative research was based on peer-to-peer, individual relationships, and it's unclear whether private industry was expecting any results from collaboration. The amount involved is too small; a lot of donations are less than ten thousand dollars. This, from firms' perspective, probably means that they want to maintain good relationships with specific professors, want to gain access to promising graduates, or want to run very small collaborative projects. These are great things; we would like to encourage them.

In this account, the enforcer of the university rules about formal collaboration stated that he would encourage donations (and the research associated with them); he excused them as very small and probably not likely to produce substantial results. As we have already seen earlier, these are precisely the characteristics of informal collaborations: a relatively small amount of donations for recurrent years, with the expectation that the relationship will be long-lasting. But this administrator's account neglects the rest of the informal bargain: if inventions do emerge from a collaboration, firms expect the professor to give them the IP rights, with the understanding that the firm will reciprocate later on. His account also obscures the fact this approach does not compensate the university as the formal procedures do. Essentially, he is admitting that the donation system is, to some extent, here to stay, although he hopes that large-scale collaborations will be handled formally through the university.

Sticky Institutions

The "good old days" had to change, but their practices did not disappear after the Japanese Bayh-Dole Act and related measures were implemented. Curiously, it seems that there wasn't anyone who wanted the old practices to disappear—the laws, the rules, their implementations, university administrators, and university scientists all contributed to what new institutional theory calls the decoupling or loose coupling of formal rules and the associated practices. The theory predicts the loose coupling, but it does not predict how it takes place. But unraveling the process through which the new rules were created, implemented, and modified makes it clear that the loose coupling is an accomplishment. University professors, enmeshed in the meanings and practices associated with the collaborative relationships with firms they had fostered since the 1970s, ignored the rules when they could and willingly perpetuated an arrangement that enabled them to give firms "favors"—often in the form of IP rights. But scientists were not responding to the new rules in a vacuum, and this case of loose coupling was not accomplished solely by their individual behavior. They responded to the new rules creatively, based on careful consideration of what the rules and the university administration would allow, observation of what kinds of nonadherent actions would be considered within the realm of legitimacy, and regard for what made sense to them as university academics. In the end, they were the ones who created the decoupled actions, but their choices were scaffolded by the new policy and its actual implementation at the level of policy and organizations.

The loose coupling, in this sense, started from the very beginning, at the moment the Japanese government started to consider introducing a new, more formal structure for university research commercialization by imitating the rules in the United States. The U.S. structure for research commercialization had arisen from the practices of a select few private domestic universities. Transplanting this structure into the context of Japan's national university system meant that the universities needed new formal organizations, funding, and government guidance for IP management. From a policy perspective, university-firm co-patenting was a good solution to these problems, especially that of funding. By offloading some costs associated with patenting to firms, the national universities could start obtaining patents at low cost. Although co-patenting meant that universities' IP rights would be shared with firms, the patents still counted fully as university patents, which was bureaucratically important. From the firm's perspective, being able to secure IP rights to any collaboration with a university professor was simply "the way it should be"—the way university professors and firms had operated ever since university-firm collaboration had existed in Japanese academia. Firms had little reason to agree to universities taking sole ownership of patents, so they settled for university-firm co-patenting—although sometimes bitterly because, depending on their contracts with universities, they now had to acknowledge the university's share by compensating for nonuse.

Ceremonial engagement with the formal rules and a lack of surveillance by the administration were two necessary ingredients in the decoupling of local activities from the form.[9] The case of university research commercialization in Japan shares these two ingredients. But this decoupling was also a result of loose coupling in each layer between the national legal framework, the institutions of Japanese national universities, university administrators, and, lastly, individual professors. Empowered by universities' interpretation and implementation of the policy and administrations' lax surveillance, professors created a new set of practices that pulled the new structure toward the old practices in two ways. First, they retained the informal practices as a "parallel world" where legality is dubious but scientists could keep their promises. Second, they shaped how the new formalized methods of commercialization were understood and practiced by filtering them through the lens of the old practices. The "pull" then became permanent in the sense that it shaped what kinds of loosely coupled practices were created, routinized, and then taken for granted in the new context. The final result was the institutionalization of the loosely coupled practices.

Among the new practices, university-firm co-patenting is an especially clear case of institutionalization. By 2014—roughly a decade after the implementation of the U.S. model—Japanese universities typically agreed to co-ownership of IP rights with firms that collaborated with university scientists as a standard "bonus" for such firms, provided they bear the cost for patent application and maintenance. Faced with similar pragmatic problems, all Japanese universities adopted similar loosely coupled practices, even though they hadn't been instructed to create such a specific loosely coupled form.[10]

Japanese policymakers and university administrators did not seem troubled by this development. In fact, a Tokyo University report commissioned by MEXT acknowledges "the spirit of Wa (peace, cooperation), skills at collaborative work, [and a] high-ratio of co-application [of patents]" as Japan-specific conditions of university research commercialization.[11] In the 2010s, the compensation for nonuse clause became the focus of discussions within universities and MEXT, indicating that co-patenting had become "the normal way" to patent faculty inventions.[12]

The case of research commercialization in Japan suggests that it's inaccurate to think of decoupling as a side effect of institutionalization. Instead, I argue, decoupling is *a part of* institutionalization in the sense that the "pull" of prior practices affects how new rules are implemented. Variation in a diffused global form will persist to the extent that loosely coupled forms are institutionalized locally and variations do not cause problems between localities. The institutionalization of research commercialization in the United States was a historically specific construct that depended on the interactive responses of scientists, policymakers, industry leaders, and university administrators.[13] The imitation of such a construct is also historically contingent. Through the process of importing, translating, modifying, partially ignoring, and reinterpreting, every actor involved—ranging from policymakers to scientists—contributed to the loose coupling between the original form in the U.S. and the bricolage of new practices in Japan.

FIVE Institutional Travelers

Japanese University Scientists as Cosmopolitans

Professor: I'm a founding scientist of [an American bioventure]. Now we have one drug with FDA approval, so we'll be speeding up. [continues to explain the details]

Interviewer: How did you meet the people in the bioventure?

Professor: They just came all of a sudden in 2001 to our lab; that was the beginning of our relationship. From our perspective, it's like, we did collaborative research with an American firm and the FDA approved it, then the PMDA [the Japanese FDA] would say, "Oh, if the FDA approved it, we will too." Isn't this pathetic? But I'd rather go the United States route than try to change the system. [. . .] Researchers are a flexible species, in terms of this, so we can weave our way through it by international collaborative research.

These quotes, from a medical school professor at a prestigious private university, represent the ease with which elite scientists discussed the topic of university research commercialization with me. They intuitively understood my research question—how do we understand the adoption of U.S.-originated commercialization policies in the Japanese context? What's more, they spoke with ease and familiarity about science and commercialization in the U.S. and other countries, even when they themselves had never been involved in international commercialization endeavors. In fact, the medical school professor quoted previously had pursued much of his commercialization in the United States.

He had first gained access to the U.S. academic and commercialization world when he lived in the United States for two years as a postdoctoral researcher, and even though he then moved back to Japan at this early career stage, his ties with the U.S. commercialization world grew stronger, not weaker. At the time of our interview, which was more than a decade after his return to Japan, he was running several international collaboration projects that included sending some of his PhD students to the U.S., in addition to being involved in the U.S. startup he mentioned. The IPs for his main drug candidate were exclusively licensed through his Japanese university to the American startup. At the same time, he was working with two large Japanese pharmaceutical companies on different projects under collaborative research contracts. In Japan, he uses only the new formalized methods of commercialization, and his university is the single assignee of all his inventions. But that does not prevent him from commercializing his inventions elsewhere around the globe.

"Researchers are flexible species," he told me—and his experience demonstrates that. Collaborating internationally and having a startup in the U.S. had enabled him to ease his way through the Japanese drug administration by first obtaining approval from the FDA. In fact, bypassing the Japanese drug approval process was a common practice for Japanese university scientists. As elite, successful bioscientists, they belonged to two research commercialization institutions: the Japanese and the U.S. systems. They thought and acted as cosmopolitans who were not necessarily bound by the regulations and practices of Japanese academia.

This chapter describes the commercialization practices of these university scientists who have experiences and resources outside Japanese academia. The scientific community is a global one, so most Japanese university bioscientists inevitably have extensive communications with the global scientific community. Their academic achievement is measured by publications in international peer-reviewed journals such as *Science* or *Cell*, they have attended and organized professional conferences worldwide, and they collaborate with colleagues internationally. Among the scientists I spoke with, international collaborations ranged from the trivial (exchanging mice for experiments) to running a large international project with multiple labs and grants. Furthermore, because postdoctoral training in the United States (or, sometimes, Europe) is a typical trajectory for promising Japanese bioscientists before they assume positions in Japan, over half my interviewees had spent one to five years abroad. Some had held faculty positions in the United States for many years before returning to Japan.

To capture the agency of these cosmopolitan scientists, I introduce a new concept: *the institutional traveler*. This concept describes the theoretical link between these professionals' embodied experiences of interacting with scientists and firms in the U.S. and Europe, which influenced how they devised their commercialization activities in Japan before and after the policy changes. The institutional traveler concept elucidates the continuous and continuing effect of a globalized world where many professionals travel to different institutional settings that feature different rules, cultural assumptions, and resource constraints. As inhabited institutionalism shows, actors creatively use local and extralocal meaning systems to negotiate new sets of practices.[1] For example, as Chad Michael McPherson and Michael Sauder show, a state attorney in a drug court can, in some situations, evoke the logic of rehabilitation to negotiate for a commuted sentence, even though such a logic institutionally belongs to the psychologists, not to state attorneys.[2]

But how creatively can an actor in an institution behave? How extralocal can their source of inspiration be? Japanese university scientists, it turns out, can create new practices by synergizing material and cultural resources both locally and extralocally, beyond national boundaries. The scientific community is global and cosmopolitan, and elite Japanese university scientists who work in it are institutional travelers. When the professor quoted at the beginning of this chapter said "researchers are a flexible species," he was speaking about the cosmopolitanism of belonging to Japanese academia but having gained knowledge, experiences, and resources from the global scientific community.

Knowing the Global Rules in Advance

The "advanced" world of commercialization was within the realm of these scientists' scientific life well before the U.S.-originated policies on research commercialization began to diffuse around the world. Interaction with the global scientific community made scientists realize how distinct the situation in Japan was, especially in contrast to the United States, where academic entrepreneurship and formalized methods of commercialization have been practiced since the 1980s. Japanese university scientists understood commercialization practices in Japan as a specific, contingent, and "Japanese" way of doing things that needed some explaining. A pharmacology professor at one of the best universities in Japan elaborated on this point:

> I didn't have any business [with U.S. firms] but had networks with them on a personal level. Some [American and European] university professors I knew had also joined companies, and you'd hang out with them. But I think they also had this understanding that Japanese university professors have certain ways of having relationships with firms. It might depend on your specialties, but at that time [1980s–1990s] I used to explain to foreigners, "This is the way Japanese national universities work, the way Japanese professors work with firms." Because they were changing after the Bayh-Dole Act, but we had nothing like that.

Another senior scientist in plant biotechnology recalled his interactions with American firms in the 1980s:

> Scientist: My professor talked about my dissertation work at a conference in the States, and [a major American pharmaceutical firm] got interested, and three, four people flew to Japan to talk to me. I was very surprised; that was the first time [I had interaction with U.S. firms]. It's the methodology I'm still developing, on antibodies. I thought, "American people have different ways of thinking!"
>
> Interviewer: How come?
>
> Scientist: The concept was brilliant, but practical application would have taken time—and they just funded me for two or three years, just like that.

Whether or not they went on to participate in the "American" system, both interviewees were well aware of what that system of commercialization entailed. The pharmacology professor had not engaged in active commercialization; he was exposed to the increasingly aggressive nature of commercialization in the U.S. and Europe indirectly, through the eyes of his colleagues in the United States and beyond. The plant biotechnology scientist recalled that as a new PhD, he had been surprised to find out how aggressively American firms sought to find out about and fund his research.

As these examples illustrate, top scientists in Japanese academia inevitably participated in North American and other international scientific scenes as part of their everyday research activities. The "taken-for-grantedness" of local arrangements that is often discussed in the literature of institutionalization and diffusion was notably absent for these scientists.[3] Even those who had relatively limited direct experience with U.S. scientists communicated with their global peers frequently, and in their interviews they talked about "professors with a yacht and Porsche in the United States."

Learning "American" Practices and Applying Them Flexibly in the Japanese Context

Some of the professors I spoke with had lived in the U.S. and held an academic position there for more than a year. This sustained experience had given them a more intimate understanding of how commercialization works in the United States, and when they got back to Japan, they adopted similar practices even though Japanese academia did not yet have the same kind of scaffolding for commercialization. Without that scaffolding, they could not simply switch to the "American way"; nor was it practicable to create a compromise practice that fell somewhere between the old Japanese practices and the American ones. Instead, they innovated by negotiating new arrangements that addressed their commercialization interests within the constraints of the Japanese commercialization environment.

One example is an immunology professor who had been unusually entrepreneurial: he had included himself as an assignee of his inventions even in the early 1980s, before the policy changes. He had negotiated with firms over the terms of collaboration to ensure that he would be at least a co-assignee of his inventions so he would retain formal rights to compensation. He explained why he had been savvy enough to negotiate with firms:

> Well, once I was working with a startup in the U.S. [while in Japan], and we had disputes over intellectual property rights. They were pretty unscrupulous. We co-authored academic publications, but they completely neglected us in patents. We got into disputes and had quite a few negotiations, but it ended up that we would have had to prepare an account of expenses if we had sued. . . . I brought the idea and reagents; they had a technique to find the [ligand of a neurological receptor]. I knew if we looked for it in such and such ways, we would find it. They had the technique, so we asked if they

> were interested in finding it with us. The idea was completely ours, but they submitted a patent application without our knowledge. So they are not like in Japan, not gentlemen-like. I was naïve about business, and I wanted the science to move quickly, but it was a bitter failure.

Ironically, it was his trouble with an American firm that motivated him to adopt the "American" practice of claiming his own patent rights. His experience with a U.S. firm that neglected his contribution to the intellectual property had made him cautious of trusting firms. This experience had led him to reflect on how he should handle working with firms in general. Although he claimed that the reason for his "bitter failure" was that the U.S. firm was not as "gentleman-like" as Japanese firms, he then brought American practices into the Japanese context. After this experience, he continued to use the informal donation scheme to work with Japanese pharmaceutical firms, since that was the most practical option then, but he always negotiated to secure some rights to any inventions that arose from collaborative research.[4]

This professor was negotiating to retain his IP rights long before the Japanese Bayh-Dole Act was instituted, during a time when it was customary for professors to waive patent rights in expectation of donations later on. His patenting records show that his repertoire of commercialization was not limited to negotiating joint intellectual property rights, however; he also maintained informal collaborations with Japanese firms. He had more than thirty U.S. patents: 47% had a Japanese firm as the single assignee, 37% had a firm and himself as co-assignees, and 11% had his university as the single assignee. Even after 2004, when the transition to the formal Japanese system for commercialization was completed, only about 30% of the dozens of patents of his inventions had a university as the assignee. As this diversity indicates, his mastery of the American system did not mean that he negated or abandoned the Japanese informal practices. In fact, even after the changes, and despite the government's new official position promoting university patenting and patent licensing, he kept giving some of his patents to Japanese pharmaceutical firms with which he had long-standing relationships.

The experience of a biochemistry professor further shows how exposure to American resources and methods of commercialization led scientists to embrace more assertive, "American" commercialization while still appreciating and using the Japanese practices as well. This professor was a systems biologist with publications in *Nature* and other outlets and had spent five years as a postdoc at a high-prestige university in the United States. In Japan, he remained very active in commercialization and entrepreneurship after taking a position as an associate (and then full) professor at one of the best national universities. In our interview, he told me how his experience in the U.S. had shaped his orientation to commercialization. Speaking of his postdoctoral years in the U.S., he recalled, "The funding from the NIH started to run out, so the professor I was working with told me to stop the experiments. I went to another professor who was involved with [a startup], and they then funded me. . . . I also started a firm meanwhile. I was like a small business owner." This scientist's familiarity with the business world was grounded in these maneuvers in the United States, where faculty members were directly involved in commercialization, and in consulting jobs he had juggled during his postdoctoral years. Now having been back in Japan for over ten years, he both interacted with Japanese firms and ran his own biomedical startup for therapeutics, which he had established in the early 2000s. He applied the practices he had learned in the U.S. to the Japanese context and produced a result that is more advantageous for faculty entrepreneurship than either the U.S. or the traditional Japanese approach: the rights to intellectual property that is used to make the core products of his own therapeutics startup are co-assigned to his university and his own firm. As a result, his firm has de facto exclusive rights to use his inventions and pays only a small "compensation for nonuse" fee to his university. This arrangement would be very unlikely in the United States, where universities tend to strictly enforce their rights to employee inventions and uphold stricter norms about conflicts of interest.[5] Having learned how to be "like a small business owner" in the U.S., this scientist negotiated the intellectual property arrangements for his inventions in a way that was advantageous for his entrepreneurship.[6]

A superficial impression of this professor's trajectory would be that he simply introduced an American repertoire into the Japan context, but a closer look at his narrative in the interview and his patenting history reveals a different picture. His patenting records show that he patented twenty-five inventions between the mid-1990s and 2015 in addition to those from his prior postdoctoral research, which have American universities as assignees. Sixteen of these patents are assigned to a firm, six are jointly assigned to a firm and a university, and only two are assigned exclusively to a university. What's more, five out of the six joint patents are assigned to his startup and the university. Thus, while he was proactive about making sure that he and his firm acquired intellectual property rights, he let the majority of his inventions be assigned to collaborating firms. When I asked whether he thought that arrangement was detrimental to his interests as a scientist, his answer was no: "People are sensitive to what is valuable. If someone says [that the Japanese donation scheme is disadvantageous to scientists], you should take it with a grain of salt. Rather, the Japanese and American systems are completely different, and in Japan, intellectual property systematically flows from universities to large companies, and they feed back stable support, though it might not be obvious. It's a proper win-win relationship." This professor has some inventions whose patents are wholly owned by large Japanese pharmaceutical companies, and some of these inventions were patented after 2004. His multiple labs, in return, receive generous funding from those companies, to the extent that one of his labs is actually named after one of the pharmaceutical companies. Like other interviewees, he used the new formal "American" system of commercialization in Japan alongside the informal "Japanese" system. He maintained some established practices, such as working with Japanese firms informally and giving them many of his intellectual property rights, while he creatively worked to arrange intellectual property rights for his startup. In essence, his practices were a result of the juxtaposition of American and Japanese commercialization practices.

These scientists' repertoires of commercialization *in Japan* were inspired by their experiences in the U.S., but they flexibly applied the U.S. practices and devised arrangements that had not previously been practiced. Incorporating the "American" approach to commercialization was not about attempting to imitate it wholesale; it was about creatively constructing a bricolage of different modes of commercialization.

Another professor's experience further clarifies this point. The professor is an immune pathologist at one of the most highly ranked medical schools in Japan. After completing his MD and PhD in the late 1970s, he flew to the United States, where he worked for almost two decades, first as a postdoc and then as an assistant and associate professor, and participated in multiple collaborative research projects with both American and Japanese firms. He took a position in Japan when the policy and funding situation for commercialization started to change in the early 2000s. He recalled his early days in Japan when he was maneuvering different modes of commercialization.

> I put everything in [a kind of antibody], so I talked to all the pharmaceuticals in Japan, like [A] and [B] for collaboration, but all they could say was

> "This is interesting," not "Yes" or "No." If they say yes and fail, they'll be criticized, but they really didn't want competitors to commercialize either. I tried that, but nobody bought in, so I created a bio startup. . . . Now we're in the process of phase 2 [of drug approval] in [a European country]. The crucial thing is having good materials. Some patents are from [an American medical school he had worked at], but the humanized antibody itself was made here, and I got lucky—it was before the university got corporation status, so the university did not claim its rights, so my firm owns the patent. This drug could directly channel profit, like Dr. Kishimoto at Osaka University. If it were through the university, they would have taken a large cut. My startup got all the IP rights.

Here, as in the previous example, the professor describes introducing the entrepreneurial orientations and practices he had learned from his experience in the United States when he moved back to Japan. Unable to find satisfactory collaboration deals with Japanese pharmaceutical companies, he started his own company. And yet, precisely because Japan was only beginning to implement the new U.S.-inspired commercialization processes, he was able to strategically bypass them. He says he got "lucky" because, unlike in the United States, where the university took the IP rights to his inventions, in Japan he was able to get some of the IP rights assigned to his firm. He refers to Dr. Kishimoto, whose inventions led to an immunotherapeutic drug that generated about $730 million in 2014, and says that his own inventions might similarly lead to substantial profit that will belong entirely to his startup.

These cases exemplify how interacting with U.S. firms and colleagues shaped Japanese bioscientists' orientations to commercialization. Constant interaction with a global scientific community, especially in the United States, prepared Japanese scientists for the government-initiated introduction of the U.S. structure of commercialization. But it prepared them in a peculiar way. The Japanese environment before the policy changes—that of informal, gift-exchange interactions with firms—did not necessarily present a barrier to their commercialization ambitions, and they did not seek to completely replace Japanese practices with U.S. practices. Instead, they created a bricolage of practices that included arrangements that were not possible in the U.S., such as inventor ownership of the IP, giving IPs to collaborating firms in exchange for donations, university-firm co-ownership, and even solo ownership of IP rights by the scientist's startup.

Here and There: Two Entrepreneurial Worlds

For institutional travelers who move around in the global social world, resources themselves may be translocal. Many of the scientists I interviewed were "returnees" from U.S. academic institutions and had worked in the U.S. commercialization world for periods ranging from a year of postdoctoral research to decades of professorship in the United States (and in some cases, Europe).[7] These scientists often maintained ongoing collaborations with U.S. colleagues. As we have seen, these interactions with U.S. colleagues afforded them knowledge about U.S. commercialization practices and enabled them to create their own commercialization practices in Japan. Their time in the U.S. also gave these scientists ties to the more resourceful, or differently resourceful, world of American commercialization. A leading scientist in dermatology who seemed to be completely enmeshed in the Japanese commercialization system provides a striking example of translocal commercialization. In the late 1970s, after obtaining his MD and PhD at a Japanese national university, he did four years of postdoctoral research in the U.S., first at an American public health research institute and then at a prominent private medical school. At the time of our interview, he had been a faculty member in the medical school of a private university in Japan for over forty years. He worked with multiple Japanese pharmaceutical firms and had visitors and researchers from these firms in his labs all the time.

A close look at his IP records as well as his in-depth interview, however, reveals that he is far from following the traditional route of working with firms without seeking compensation and is adept at using different repertoires of action to pursue his entrepreneurial goals. To begin with, he has eight U.S. patents that were granted between 1990 and 2005 and is the single assignee for three of them. He proactively sought and successfully gained more control over his inventions rather than simply giving IP rights to firms. In addition, after speaking at length about his industry collaborations and inventions, he mentioned that he was involved with a biotechnology startup in the United States. I asked him to explain how he ended up having the startup in the U.S., and he recounted,

> I moved from [American public research organization] to [American private medical school], and I had a Taiwanese friend while I was working there. He said to me, "I wouldn't be able to make it as a scientist; I'm starting a reagent

> firm, can you help me?" So I gave him monoclonal antibodies, even the ones I developed in Japan. [He goes on to explain his ties to other U.S. firms]. We manufacture 80 to 90% of the antibodies from my lab in the U.S. It's cheaper to make it there, more reliable, and the market's ability is very high. [. . .] We have a few antibodies being manufactured in Japan, with [a Japanese pharmaceutical firm], but this one is no big deal. [. . .] [He explains a new commercialization candidate he is developing in the U.S.] This antibody is really attracting attention from scientists all over the world, and there is a collaborative research offer from [renowned American hospital]. We are also mass-manufacturing this in the U.S.

This scientist states that while some of the antibodies he has invented are produced in Japan by a Japanese firm, the major ones are being produced at a startup in the United States, because his ties to the U.S. provide easy access to commercialization there, which he felt was a better environment for commercialization. His fluency in Japanese informal commercialization practices does not, therefore, mean that his repertoire was limited to these practices.

His mastery at switching between the commercialization worlds of Japan and the United States afforded him the ability to transfer technology successfully across the two worlds. His involvement with U.S. academic entrepreneurship thus did not make him more "American." His commercialization repertoire combined Japan's informal practices with his own assertive control over IP rights and his experience of entrepreneurship in the U.S. biotechnology field. Institutional travelers like him enjoyed multiple opportunities to commercialize their research and multiple ways of doing so, even long after their stay in the United States had ended. Scientists with high-profile research get countless offers to work with colleagues or firms internationally, as exemplified in this chapter's opening story about an unexpected visit from a U.S. startup. These elite Japanese scientists' interactions and work with international colleagues and firms give them a translocal view of commercialization and make them, in the words of that opening excerpt, "a flexible species." That scientist's cosmopolitanism and lack of attachment to the Japanese commercialization world are clear in his remarks. International collaborations enabled him to ease his way through the Japanese drug administration by first obtaining approval from the FDA. In fact, bypassing the Japanese drug approval process was a common practice for Japanese university scientists who are also entrepreneurs.

All these scientists, including those who seemed to prefer "local" arrangements, were able to bypass part of the Japanese commercialization world when an alternative environment that they had access to seemed like a better option. They did not confine themselves to one site for their scientific and commercialization efforts; they created multiple pathways to commercialization that they used on a case-by-case basis.

The Social World of Cosmopolitan Scientists

The elite Japanese university bioscientists in this chapter demonstrated a certain ease in being institutional travelers and taking advantage of the two worlds they belonged to. They expressed a sense of gratification and pride at having managed to weave through different environments to become successful academic entrepreneurs. Their experience highlights the analytic leverage of Alvin W. Gouldner's notion of cosmopolitan actors, which is largely forgotten today.[8] Gouldner was interested in latent social roles—identities that are less relevant to the social situations at hand but nevertheless exert influence.[9] To understand these identities, he surveyed university professors about their loyalty to their university, their commitment to professional skills and values, and their reference groups. His insight was that while a professor's manifest social role was that of an educator in a university, many "cosmopolitan" scientists were more committed to their research and professional communities and sought reference groups that included other renowned researchers outside their university.

Gouldner conceptualized cosmopolitan scientists as those who affiliated with the larger U.S. professional community, but his concept clearly resonates with elite Japanese university scientists' orientation toward research commercialization. Their reference groups are the global scientific community that transcends national boundaries, and they also have access to resources outside of Japan, often including venture capital in the U.S. The notion that researchers are a "flexible species" connotes some degree of detachment from local-national constraints. Being an elite scientist in Japan meant being embedded in a global professional community; having access to scholars, firms, and other resources in the U.S. and elsewhere; and feeling a sense of belonging outside Japan as well as in Japan.

It would be wrong, however, to argue that local-organizational rules and values do not matter to cosmopolitans. Rather than thinking of cosmopolitans as a special group, Barney G. Glaser argued—in response to Gouldner—

that localism and cosmopolitanism should be thought of as dual, potentially conflicting orientations of highly motivated scientists.[10] The cosmopolitan scientists in this chapter worked with their local-organizational constraints, just as the cosmopolitan professors Gouldner studied cared about teaching and university administration in addition to their research. The prominent scientists interviewed in this chapter still gave many of their patents to the Japanese firms they collaborated with, even though that arrangement risked breaching the rules and rested on trust rather than formal agreements. The fact that it "made sense" to give firms their IP rights attests to the significance of the established Japanese practices and the expectations of these scientists. Being cosmopolitan, then, does not mean attending only to the global professional community and its resources, and it doesn't even mean having a bigger repertoire of commercialization strategies. Rather, it means traveling over time through institutions in different contexts. Cosmopolitan Japanese scientists are bound and constrained by multiple different frameworks, but they also have multiple ways of solving the problems associated with those constraints because of their position in Japanese and international academia.

Elite Japanese bioscientists' juxtaposition of commercialization practices reveals what theories of globalization, translation, and the diffusion of policy have often missed: the global flow of people, which is interdependent with but somewhat autonomous from the global flow of ideas. Institutional travelers further complicate the story of how new practices emerge. The institutional travelers in this chapter applied elements of the new forms of commercialization even before the U.S. policy diffused to Japan. They selected from among the previous informal practices and the new formalized ones, depending on the situation, and took advantage of their ties to the wider biotechnology world. Globalization, then, is shaped at the interstices between the diffusion of an institutional form and the cosmopolitan actors who travel through multiple institutions over the course of their careers.

In the case of commercialization practices, this means that the new institutionalized practices must be understood not as a compromise or an inferior variation of the model practices that the new rules suggest. In the previous chapter, we saw how the gift-exchange-like tradition still significantly influenced how scientists chose to commercialize their research. Scientists loosely coupled their practices with the rules because the previous gift-exchange-like practices facilitated their relationships with firms. Meanwhile, the new practice of university-firm co-patenting—which gave the firm privileges like those they were accustomed to—became the new "default." The decoupling was also shaped by the repertoire of innovative practices the cosmopolitan institutional travelers created: they simultaneously used the formal "new" commercialization practice that promoted university ownership of patents and licensing, the traditional gift-giving practices that granted firms IP ownership, and an even more proactive strategy that retained IP rights for themselves or their startups.

Taking the last chapters together, we start to see a holistic picture of how U.S. commercialization rules were adopted and adapted in the Japanese context. The old gift-exchange-like practices exert a pulling effect on the new rules and create loosely coupled practices that are distinct from the U.S. commercialization practices, and institutional travelers create further variations. The new practices emerge as each scientist finds a way to commercialize research in a given situation. And as practices accumulate, divergent but patterned responses to the rules become the new "way things are" in Japanese academia. The resulting picture of Japanese university bioscientists' research commercialization is a multifaceted, agentic, and creative one featuring practices that differ from both the old informal practices and U.S. practices.

Japanese scientists' responses to the new policies demonstrate familiarity with the global scientific community combined with a clear understanding of their position in the Japanese scientific community. In fact, all the scientists in this study were acutely aware of being Japanese, including working in Japanese academia and identifying as culturally and ethnically Japanese. The next chapter fills in this final piece of the picture. How, it asks, did Japanese university bioscientists understand their changing commercialization practices? How did that understanding shape their sense of themselves as scientists? And how did they come to think of their practices as natural, sensible, meaningful, and normal?

SIX A "Japanese" Collaboration

IN THE UNITED STATES, the Bayh-Dole Act was framed as an important way for the U.S. to compete against the rising economic powers of Europe and Japan after World War II. It was passed during a period of economic stagnation with the objective of regaining the country's competitive edge by fostering innovation. This effort to stay competitive had circular effects. When the Bayh-Dole Act and the rest of the U.S.-born structure of research commercialization spread abroad, countries around the world framed their adoption of it by referencing the United States and its success as a model for their own efforts to stay competitive in the global economy. In fact, the Ministry of International Trade and Industry's explanation of the Japanese Bayh-Dole Act (Industry Technology Enhancement Act, Clause 17) starts with a reference to the U.S. context: "Background: Because of the decrease in the international competitiveness of the American economy in the latter half of the 1970s, the United States passed the Bayh-Dole Act, which essentially attributes the intellectual property rights of innovation based on government-sponsored R&D to entities such as private firms. It is said that this helped American industry regain competitiveness by propelling firms' technology development and creating startups."[1] The document goes on to explain that these events prompted the Japanese government to initiate a similar act to foster innovation.

The Bayh-Dole Act and other American innovation policies are textbook examples of what Paul J. DiMaggio and Walter W. Powell term *institutional isomorphism*; under uncertain conditions, DiMaggio and Powell argue, competition encourages organizations to adopt a form that has been successful for other organizations.[2] What is curious about Japan's imitation of the Bayh-Dole Act, however, is that this conscious importation seemed to actively contribute to Japan's acute sense of locality. Policymakers talked about the "Japanese" version of the commercialization structure from the moment of adoption as a potentially improved—or, at least, workable—version of the American model. A 2010 policy paper by MEXT, administered by the University of Tokyo Division of University-Industry Cooperation, set forth a series of proposals for how collaborative research between university scientists and firms should proceed. Its first bullet point argues, "We need to start a discussion about establishing a 'Japanese form' of university-firm collaboration that is not American or European, so that the national power of Japan will grow." The document goes on to list the characteristics of Japanese university-firm collaborations, including "the spirit of *Wa*, being good at working together, a high rate of co-patent applications—collaborative research is the pillar of university-firm cooperation."[3] In these excerpts, the report compares and contrasts the (implicitly idealized) Japanese form of university-firm linkage with those of the U.S. and Europe. It indicates the relationship between the "Japanese" form of commercialization and the "essential" features of Japanese society by citing "the spirit of *Wa*"—a Japanese word for peace or order that has been used since the seventh century.

This chapter examines Japanese accounts of this process of nationalizing the adopted structure for research commercialization. What practices facilitated this national and ethnic identification process, and what role did national identity play in the adoption process? Policymakers were not the only ones to emphasize Japanese-ness when talking about research commercialization. Almost all my interviewees repeatedly described some practices as *Japanese*. Scientists evoked "Japanese-ness" by referring to policies, practices, ideas, or values as "Japanese." They referenced institutional arrangements such as national universities and commercialization policies, popular notions of Japanese culture and history, and the discourse of *nihonjinron*, a theory of Japanese-ness that is popular in Japan, to claim that their situations and actions were specifically "Japanese."

In doing "being Japanese,"[4] the Japanese university scientists accomplished an ethnomethodology of nationalism—the enaction of their national identity; that is, they became Japanese precisely because and to the extent that they managed to keep claiming that they were Japanese. In this chapter, I delineate the everyday discursive tools they used to posit that the cause of the phenomenon or action at hand derived from some kind of ethno-national distinctiveness; among these tools were lay categories of "culture," ethno-psychology, and readily available tropes about Japanese history. By making the new practices "Japanese" and therefore "ours," these scientists took steps toward coming to terms with and normalizing the new and originally foreign rules of commercialization. By zeroing in on the accounts of those who enact the adopted policies, this chapter shows how people who are tasked with incorporating new top-down, foreign rules into their local practices make those new rules meaningful. In this case, nationalizing the new practices was particularly helpful: thanks to scientists' claims that the new policies and practices were undoubtedly Japanese, those policies and practices became locally meaningful to the point that they could potentially become as taken-for-granted as the old practices had been.

Japan as a Nation-State

In interviews, Japanese university bioscientists frequently referred to Japan as a demarcated nation-state with its own policy, legal, and resource environment. They typically compared this environment with that of the United States, drawing on either their own personal experiences or more generalized knowledge about how research commercialization worked in the United States. In some cases, the comparison country was Britain, Germany, France, Switzerland, South Korea, or Singapore. Scientists were likely primed to make such comparisons by my interview questions, which asked how they commercialized their research after the Japanese Bayh-Dole Act and related measures were implemented. To explain how commercialization practices differed, or still differ, between Japan and the U.S. or European countries, they referenced these nations' macrostructural differences. For example, a scientist who had served on a government advisory committee to promote commercialization said he had made a conscious effort to imitate the U.S. commercialization environment. He described Japan's system of national universities as a legal barrier to such imitation:

> Right then the Science and Technology Basic Law passed [1995]. . . . We proposed to create a new mechanism to revitalize Japanese universities, or else Japanese industry would be put on the line. [. . .] At that time, unfortunately,

> there were many differences between the U.S. and Japan. You see, the universities that would create the seeds for commercializing research are almost all national. In the United States, there are so many private universities—of course there are state universities too—so, they can change to do [commercialization]. But in Japan, they were national employees, at that time. There were legal restrictions that banned national employees from collaborating with industry.

As this interviewee recalled his active participation in policy creation, it's clear that the framework of the discussion had been clearly demarcated by national boundaries. Science is regulated by a national legal framework and organized around state-based support in the form of grants and university subsidiaries,[5] all of which feeds into what economists call a *national innovation system*.[6] Because the commercialization of research is governed by national policies and regulations, and because most university professors were attuned to such policies, it was a short step from talking about national policy to talking about "Japanese" laws, resources, and organizational conditions.

Scientists also compared Japan's resource environment with that of other countries. One biochemistry professor I interviewed had worked for about a decade at a top-twenty U.S. biochemistry department and had recently been recruited by one of Japan's best universities. He considered the U.S. to have a superior and more advanced research commercialization environment and described his sense of the differences:

> In the U.S., nobody talks about industry-university cooperation. You leave them alone, and they'll do it. [. . .] Perhaps it was hard in the United States, in the past, when the Bayh-Dole Act passed and the universities all of a sudden got tasked with selling IPs—I don't know about how it was back then. Maybe it was similar to the situation in Japan now—the economy was bad too. So Japan is behind, I would say. [. . .] As a summary, I say the difference is marketing and the buyers. Japanese pharmaceuticals are too cautious to buy a risky investment. They are risk averse too in the U.S., but there are so many venture firms, and they buy more easily. [. . .]
>
> In terms of institutions, Japan made progress, or it became similar [to the U.S.], but what can't change are marketing and the number of firms. This country doesn't have the system to pop up so many startups like the U.S. That's probably why the government has to shout "industry-university collaboration," I guess. In the U.S., a startup comes about without any intervention. Japanese government officials know this difference—Japan won't

> become like the U.S. without intervention. You see, the old man who lived across the street in [the U.S. city where he had lived] was investing in startups, and my dad is going for a postal savings account—what a difference. It's different, where they invest, even the old people.

This returnee eloquently explained his analysis of the differences between the Japanese and U.S. resource environments. In this excerpt, he correctly notes that there were numerically more firms in the U.S. that might be interested in buying IPs from a university. He also describes differences in risk-taking between Japanese and American pharmaceutical firms. He contrasts Japan and the U.S. in terms of their different resource environments for commercialization, including different organizational setups and different inclinations for risk-taking. This was the most common way scientists referred to the "nation"—as a demarcation that directly shaped the commercialization environment.

Toward the end of this excerpt, this professor hints at differences in the psychology of individual Japanese and American citizens: his father, depicted as a typical Japanese, saved his money in a postal savings account, whereas the professor's American neighbor invested in startups. He used these two individuals to represent "national" characteristics and thereby constructed the figure of "the Japanese" who invests more cautiously than the American.

Conflating Institutions, Practices, and "Japanese" Culture

Another way scientists invoked the nation was by referring to customs, institutions, or values that were widely considered "Japanese." For example, some professors characterized Japanese society as rigid; they described skepticism of novel ideas, an emphasis on trust or relationships over profits, heavily bureaucratic government organizations, and a legacy of lifelong employment. Such comments often came in tandem with descriptions of legal or institutional differences—such as differences in employment systems—that made their claims about Japanese-ness seem more credible. For example, they often cited Japan's tradition of lifelong employment with the same firm as a reason why startups and other risk-taking behaviors are unlikely to succeed in Japan. An assistant professor of bioengineering described the challenges associated with doing applied research and needing a postdoc to engage in collaboration with a firm: "It is difficult. It's Japanese society, so you can't take chances that may become trouble later, if you hire a postdoc. Once you hire a person, you'd try to find him a position [after the project ends], and what if he didn't have publications? Then you wouldn't be able to find a position. Japanese society is not fluid." In this excerpt, the professor describes how characteristics of Japanese society shape his scientific life. This professor worked at a national university, where, at the time of the interview (2017), it was technically completely possible to hire a project-based postdoctoral researcher—but he hesitated. As in the U.S., good academic positions are scarce in Japan, and postdocs in general find it difficult to land a good academic job after their contract ends. But in this professor's eyes, "Japanese society" would oblige him to find the postdoc a position after the end of the contract, which made him reluctant to hire a postdoc. In fact, this professor's advisor—who worked at the same university—had explicitly told him not to hire anyone unless he felt confident that he could help the person land a tenure-track job later on. The norm and the reality thus reinforce each other: this professor himself had held the recently created job position of "project-based assistant professor"—a short-term, contract-based professorship with grant money—until he "slid into," in his words, his current tenure-track assistant professor position with the help of his advisor.

Similarly, a retired professor who had become the CEO of a research tools and analysis startup described struggling with human resources management because of "Japanese" employment relations. He had been talking about the precarity of employment at U.S. startups and mentioned that his firm, by contrast, never laid off scientists due to underperformance:

> CEO: In Japan, Japanese startups don't do that. My firm also doesn't throw people out based on competition. [. . .] The employees came to work here, so we consider what is beneficial for them, too. We don't feel we're "using" them. So if someone whose ability is half of others happens to come here, the firm's output will just decline. To change the human resources, [it's like] the firm would have to go bankrupt [to rationalize] getting a new batch of people. Japan is that kind of country. So there is none of this "I was recruited from this other firm, so I'm switching"; "Oh, what was the offer? We'll offer more to retain you; please stay and work hard for us." [. . .] People say there aren't enough startups, not enough venture capitals, policy initiatives are just words, but I think one of the big reasons [why startups don't prosper in Japan] is

> that there's no understanding of using people as disposables. So in that sense, Japanese society is stable.
>
> Interviewer: You can't fire employees?
>
> CEO: No. So in my company, everyone is fine and working hard, but, and this happened last year, when an American firm comes and head-hunts our brains, we can't do anything to prevent it.

In this excerpt, the professor-turned-entrepreneur attributes Japan's small number of startups to Japanese society: in particular, employment customs made it very hard for him to reorganize human resources promptly. As the CEO, it was certainly possible for him to hire, lay off, and compensate competitively within the constraints of Japan's employment laws, which do offer more labor protections than U.S. laws offer. Nevertheless, he claimed that Japanese society—and he himself, as part of it—could not make such a leap.

In both these cases, the professors treated some characteristics of Japanese society as stable and factual and used this "objective reality" to explain their own practices. Yet such nationalizing accounts were not obvious. In reference to the first case, postdoctoral positions are fixed-term employment, so there is no legal obligation for the interviewee to find the postdoc a subsequent job. This professor nevertheless felt a responsibility to do so because it was customary in "Japanese society." In the second case, the CEO was indeed bound by employment laws that restrict the termination of full-time employees, but those laws are by no means Japanese historically or comparatively—in fact, they weren't enacted until the 1950s and were modeled after those of many European countries. But the CEO frames these laws and his value for treating employees as people as "Japanese" in contrast to the United States, which in this case is treated as the "quintessential West."[7] Both scientists cited the weight of tradition as the primary reason they could not act as American scientists would have.

Some nationalizing accounts were less convincing than others but did the job of accomplishing and reproducing a nationalizing narrative. An interview with a prominent immunology professor at a private university makes it evident that claims about Japan's specificity are *claims*—that is, performative acts that construct Japanese-ness. Prior to the excerpts that follow, the interview covered various topics related to this professor's research commercialization, including his career, the biotechnology and pharmaceutical firms he has collaborated with, his patenting practices, what happened to his lines of potential therapeutics, and commercialization news in Japan and the United States. I then asked whether he would ever be interested in establishing a startup. He replied, "No, but I cooperate [with firms] a lot. What good would it do if I did?" and immediately told me about a Japanese person he knew who had transformed himself from a scientist into a biotech entrepreneur in California. "So *that* was about money making," he commented. He then told me about an antibody molecule that had the potential to be used for an antibody therapy that was attracting worldwide attention. He was producing the antibody in a bio-startup in the United States. Again, he offered an account that invited listeners to understand his actions as *Japanese* as opposed to American:

> Well, I'm the one who started all of it [creating the line of antibodies]. And the firm listens to what I say; they cooperate. It's not about money; we cooperate with each other so I can establish new antibodies. It's not about each transaction—that's what I hope you understand. See, Americans would never understand this [laugh], but nevertheless we work together. But the president of the biotech firm visits Japan often and understands my intention very well. His wife is a transplant doctor; she understands the importance [of the potential therapeutics], so he understands the importance [of the antibody] as well. He used to be a researcher at [an American research hospital] but also holds a business degree; he can deal with medicine and stocks at the same time. That's the president, so he treats me as special; he cooperates well, though that's probably not surprising, since he owes most of their monoclonal antibody products to me. I hope you understand this—Japanese people's sensibility.

In this account, the scientist also claims that Japanese people collaborate to pursue goals and do not prioritize money. He means for his remarks to substantiate his claims, but the facts in his account do not align with his claim—his collaborators are Americans in an American firm, even as he claims that "Americans would never understand that." He recognizes that this contrast is untenable and so makes an exception for a particular American—the president of the firm he collaborates with, who, he says, understands the professor's "Japanese" sentiment. Even in describing this exception, little in his account entitles the collaborator to this honorary "Japanese" status except that he visits Japan regularly. The fact that the collaborator's wife is a doctor, and the collaborator's own status as both an entrepreneur and a scientist, both seem irrelevant to national or ethnic identity. So in the end, the professor finally categorizes him as an American and hints that the reason the American collaborator listens to him and agrees to the Japanese style of collaboration is that he owes many of his therapeutics to the professor's inventions.

Directly after this remark, I commented, "This [success at creating therapeutics] means your achievement as a scientist has paid off." He responded by elaborating on his sense of the contrast between his way of working and U.S. practices:

> Well, I mentor seven [or] eight young graduate students. What they see is . . . that they are learning from us. I think they would judge us as stingy if we were negotiating and getting funded by firms. After all, in order to nurture people, the Japanese people's sensibility . . . most of the students are from Japan, although we have about ten foreign visitors. . . . To nurture good Japanese researchers, I think we need some kind of moral standard. . . . By that I mean to look ahead, to invest in the future. There are many American scientists who have become very wealthy, with yachts and all—some of my American colleagues have retired early and are traveling all over the world. That could be nice, but I think like a poor person. I won't retire.

As in his previous interview excerpt, the argument is not necessarily logically consistent: his collaborator was American, he and his lab did receive substantial financial rewards, half the people in the lab were not ethnically Japanese, and he had commercialized his research in the United States as well as in Japan. Moreover, trust and long-term collaboration are among the most important aspects of successful collaborations elsewhere,[8] as is the discourse of passion for creating useful applications such as medicine.[9]

The analytic lesson here is not that the professor is imprecise; it is that, in his accounts, he was able to claim that certain values—long-lasting, nontransactional relationships; disinterest in money; cooperation—are Japanese and *that he works as, and is, Japanese.* He contrasts his American peers, who own yachts and retired early by profiting from their research, with himself, saying he won't retire because he thinks like a poor person. For "poor person," he used the word *binbounin*, which connotes both that he did not make an outlandish amount of money, unlike the American peers under comparison, and that in his character, he is diligent, hardworking, and modest.

What his claim did, in essence, was sustain the narrative of the specificity of Japan while adopting commercialization practices from the United States and leverage that narrative to justify his own practices. At the end of the interview, he concluded that he hoped that I—a Japanese co-national—understood "the spirit of Japanese-style industry collaboration," thereby enlisting me in the project of co-constructing nationalizing accounts.

So far, this chapter has discussed accounts of Japanese-ness that are based on either (1) policy or legal frameworks that are set by national boundaries or (2) institutional and cultural features that the scientists think of as Japanese. In these accounts, differential commercialization outcomes between Japan and other countries—particularly the United States—were the direct consequence of national differences in institutional settings, resources, and cultural values. Professors brought up a presumed set of Japanese cultural characteristics, such as valuing personal trust and being unmotivated by monetary incentives, to explain how and why university-industry collaboration is different in Japan than it is in the U.S. and elsewhere.

The "Japanese" Psyche and History

In some cases, scientists' institutional accounts slipped into assertions that institutional differences between Japan and other countries stem from embodied differences between Japanese people and people elsewhere—like the story about the scientist's father who saves money in a postal account. Pragmatist C. S. Peirce developed a semiotics that brings the effects of meaning-making to the analytic attention. Following Peirce, semiotics scholar Richard J. Parmentier calls the semiotic move to generalize an instance "upshifting"—it makes an instance a case of something larger, more systematic, and "deeper."[10] Upshifting was usually accomplished either by referring to analogous cases in Japanese history, especially its early modern to modern times, or by referring to deep psychological traits that the scientist implied were typically Japanese.

In some cases, some well-known *nihonjinron* seemed to be at work, though implicitly. *Nihonjinron* is a theory of Japanese-ness developed by Japanese academics and public intellectuals in the 1970s and '80s.[11] One prominent example of popular *nihonjinron* is the argument that Japanese society fosters feelings of dependency and the need to be accepted by others.[12] A cancer biotechnologist at one of Japan's top universities seemed to invoke this idea when he explicitly claimed that cultural differences make "Japanese" science different from science in the United States (and Britain, in this case). In a discussion of funding, he said, "The American system tries to fund young researchers with large amounts of funding. But I don't think that would work in Japan." When I asked why, he explained:

> The environment they grew up in is different. In America, they foster individualism. But in Japan, even now, it's not like that. They may be independent from their parents, but somehow they aren't standing alone from their surroundings, society. There's something Japanese, on the level of consciousness. It goes only as far as rebelling against the parents . . . but then, if you listen to the content of conversation when people gather, everyone is copying each other, although you can see the same thing in the U.S. too.
>
> So can you go independent [become an independent researcher] just because you have the skills for experiments, if you stand alone as a young researcher? [. . .] There is a joke that in Japan, they won't fund you when the answer to "Has anyone else done this research?" is no; over there [in Great Britain] they will only fund it when they know nobody has done that research. Americans are in the middle.

In this excerpt, the scientist first contends that the way funding agencies choose projects in the United States would not necessarily work in Japan. Because Japanese people are raised to be dependent and to belong to groups, he suggests, young researchers may lack the ability to start novel projects on their own. Although he was unsure whether this cultural orientation would always be detrimental to research, he was sure that there are national differences in funding preferences that are rooted in the personalities fostered by different cultural environments. Differences in funding preferences, he argued, stemmed from national differences in the consciousnesses of individual researchers—and individuals in general—in different cultural environments. Japan, he suggested, was on one end of the spectrum as the most wary of novelty, and the U.K. was on the other as the least wary of it.

When comparing Japan and the rest of the world, some scientists framed Japan as "not in the West" and also not modern. The claim was that contemporary Japanese society has a continuity with history—that contemporary issues in scientific practice have their roots in Japan's past. An emeritus professor who later became the director of a publicly funded research organization claimed that the Japanese do not possess a "modern self." He had worked at an American public research organization for ten years during the 1970s and '80s, and in our interview he mused about going back to the United States, as he is a green-card holder and his son is American. He talked about a graduate student he advised who had told him he was considering leaving academia because he wasn't excelling in his graduate program. When the professor agreed with his idea of leaving academia, the student had gotten very emotional, even though he had initiated the conversation himself. The professor had told me that it would be much easier to advise people to change careers in the United States, whereas in Japan the standard advice was to "stick with it" and changing careers was stigmatized. As another example, the professor lamented the fact that there were no direct PhD programs in Japan because of perceived equity issues:

> Professor: So once I asked [why there were no direct PhD programs in Japan] and was told that the reason is that if one person gets into a master's program and the other a PhD program, the PhD student would be preferred, and that would create awkward feelings. How stupid is this? Who cares about it? And this, similar things continue with Japanese government bureaucrats in their fifties—the one who becomes the director [of a ministry] has to be based on tenure. They still keep such customs without apology; this is exceedingly strange. So I've found so many of these issues—Japanese mass media or society—we can't correct them one by one; it's impossible. So bottom line, Japan is not suited to be a modern country.
>
> Interviewer: Did you say Japan is not suited to be a modern country?
>
> Professor: Yes. Unsuited. See, the intellectual level is high, everyone is earnest, everyone studies hard, it's all good, but something fundamental is missing. But then, everyone would complain if we were to go back to the Edo period.

In this excerpt, the scientist gives two examples of practices he deemed irrational: ensuring equality among graduate students rather than competitive selection, and prioritizing length of service when deciding who to promote. He presents these practices as evidence of a "Japanese" problem associated with prioritizing the avoidance of friction and uncritically maintaining irrational customs. He claims he's seen parallels in all parts of Japanese society and concludes that Japan is not a full-fledged modern country.

"Backwardness" was certainly a thread in these discussions. But when the scientists framed Japan and the commercialization of research in Japan as behind, not modern, or bound by customs, their reference point was the West—meaning the United States and select European countries—and with that limited sample, Japan appeared as the *only* "backward" nation. In other words, one *could* compare Japan's research practices to those of any country—

perhaps China, South Africa, or Argentina. But in practice, the comparison is always between the West, whose practices are described as modern, money-driven, rational, and efficient, and Japan, whose people are described as just not quite Western. The notion that Japan and the West are binary opposites is taken for granted, articulated, and reinforced.

The Slippery Slope of Nationalizing Accounts: Intertwining Facts and Claims for the Construction of National Identity

Nationalizing accounts are abundant in my interviews, but it's important to remember that these claims about Japanese-ness were woven into conversations that were otherwise about research commercialization. The theme of Japanese-ness, in other words, is the constant refrain, not the melody, and that contributes to its ubiquity; the idea of the Japanese nation routinely appeared throughout the conversation. Scientists shifted seamlessly from seemingly objective institutional accounts to assertions about Japanese people, culture, or history. Such changes from the institutional to the cultural and the psychological made the argument seem more natural and uncontested, as if it were simply common knowledge about Japan.

To see that fluidity in action, let's consider an extended example of a scientist who completed his PhD at the university where he now works as a successful biotechnology professor and holds a senior administrative position with the Faculty of Sciences. His career trajectory represents upward mobility in the traditional Japanese academic system. He told me that he had long been an avid reader of twentieth-century Japanese political theorists and had enjoyed his college sociology courses, especially lectures on Max Weber. He regretted never having had a position abroad like many of his colleagues; when he became an assistant professor, the position had to be filled immediately. He emphasized that he wished he'd had international experience, as he felt that his colleagues who'd had more international experience "think deeper"—"They can look at themselves from a perspective," in his words—whereas he felt that his thinking tended to focus narrowly on specific situations his lab and university were in.

This topic of constant comparison between the U.S. (and the world) and Japan created many openings for him to make claims about Japanese specificity as taken-for-granted knowledge. Here, I present a long segment of his interview to show how this recurrent theme is provoked during his narrative. We were talking about faculty members' different responses to commercialization changes. A colleague of his who had been against industry partnerships had recently begun research with a pharmaceuticals company, admitting that "the times have changed." He concluded that throughout the life sciences, university-industry collaboration had become normal. He continued his remarks and thought about how the custom of billionaires' endowments to the universities in the U.S. could be transplanted in Japan. Then the scientist expanded on the collectivism in Japan:

Professor: This may sound strange, but we always fine-tune the advantages and disadvantages of what Americans do, so, I have this discussion with my colleagues about how Japan can spin it in an original way. In Japan we can do it in specific ways, so in that sense I look forward to more successful cases in the U.S. Recently, I'm interested in the way donations are completely in place [at American universities]. [. . .] So [in the U.S.] they'd donate with the wish to help the scientist succeed. For example, you'd hear [in the U.S.], "This building of the university hospital was funded by one guy, with 50 million."

Interviewer: Right, so many named hospitals in the United States.

Professor: So these named buildings, for example, could have been "Fuji Film building" or "Pfizer building," but then that would attract a lot of criticism [in Japan]. It still does. I think it is a good idea—I think it means synergy between the society and university research. In the United States, it's named after an individual, but in Japan, it would be named after a company. In Japan, it's not about an individual but more like *Daimyo*. [*Daimyo* refers to powerful feudal lords in the Edo period (seventeenth to nineteenth centuries).] [. . .] You know, Mito-domain or *Kishu*-domain [*Daimyo*'s landholdings] were wealthy in the Edo period. It's the same sensibility. So that's how Japanese people make sense of it, like to think in that way.

Interviewer: One criticism of Japanese university-firm collaboration is that often the collaboration is done with large firms, and that does not help foster startups.

Professor: Right. In the end, as part of Japanese culture, people like intrapreneurship. [Intrapreneurship is a Japanese-English word for a startup within a large company funded by the company but otherwise detached from it]. When I'm talking to them [startups], their introduc-

> tion is like "We belong to XYZ group" or like "We are a venture firm that's closely collaborating with Pfizer." That's the difference between a culture that focuses on what the startup is or what the startup is a part of.

This extended segment exemplifies how the professors I interviewed moved seamlessly from talking about institutional differences to claiming that Americans and Japanese have fundamentally different histories and cultures. In discussing shifts in research commercialization practices, this professor brings up the theme of fine-tuning American structures to fit the Japanese context. He argues that the reasons for such fine-tuning are rooted in Japanese culture and history and that the U.S. practice of individual endowment has to be modified to a practice of donations by large firms to suit the Japanese culture of groupism. He reinforces this cultural argument with a historical example: in the Edo era (seventeenth to nineteenth centuries), the lords of large landholdings (domains) often sponsored social projects. Despite the significant differences between Japan's early modern political structure and contemporary Japan, and the fact that endowment for universities and funding for public projects are different issues, he treats these examples as instances of a continuing Japanese culture. When I point out that university labs in Japan tend to work with larger firms, he again chooses a cultural explanation rather than other possible explanations, such as a lack of "angel" investors or capital. He then seamlessly moves to yet another imputation: Japanese people think in terms of collectives, he implies, and value affiliations rather than individuals and their actions. In just a few sentences, this professor's narrative connects universities' hypothetical acceptance of large firm's donations with a historical fact about feudal lords investing in public projects on their own land and a stipulation that "as part of Japanese culture," Japanese people value affiliations with large, established organizations over entrepreneurship.

By moving seamlessly from Japan's history to its institutional environment and culture, this professor creates a "Japanese" approach to commercializing university research. He argues that research commercialization in Japan should depend on large firms because *Japanese people* prefer to depend on large organizations, both historically and as a contemporary cultural preference. These connections were not necessary; it's easy enough to explain differences between the American and Japanese institutional environments without resorting to culture. He could have argued, for example, that Japanese startups depend on large pharmaceutical firms because there is not enough venture capital to sustain them. Instead, the professor's account zeroes in on "essentially Japanese" cultural traits. This nationalizing discourse enables him to argue that the institutional differences that remain after the introduction of the U.S. scheme are legitimate and possibly even necessary for the scheme to succeed in Japan. Because Japanese people are different, he is saying, institutional arrangements should be different.

According to this interviewee, he and his colleagues observe how research commercialization works in the United States and actively modify and implement that process to suit the Japanese environment. His awareness that the legal and organizational structures were imported contributes to his pride in modifying and potentially improving their details to suit Japan's institutional environment. In his eyes, though, it isn't just institutional differences that make modifications necessary; the system also needs to be modified to *make it Japanese*, because institutional differences stem from Japan's fundamental distinctiveness rather than from the arbitrary differences one can always find between two societies.

Institutional constraints came to the fore again in an interview with a bioengineering professor who had established a startup and then left it to avoid conflicts of interest. He told me that Japanese employment customs hinder innovation, but instead of alluding to ethno-cultural differences, he used a historical narrative to nationalize the discourse. He claimed a continuity between the current state of entrepreneurship in Japan and the modernization of Japan in the early twentieth century:

> The problem in Japan [regarding startups] is that there isn't talent, because firms keep talent. In Japan, large companies hold many talented individuals and keep them there, buried. Pharmaceutical firms have lifetime employment. Ordinary people should engage in startups, right? In America, everyone gets laid off, so they come and work for a startup, but in Japan, they are retained [in large firms]. [. . .] In America, they don't keep human resources like that; they're lean, and they purge.
>
> People who remain in firms aren't stupid. Their boss may not be great, but being in the firm is pretty OK. They have a family to feed; it would be stupid to leave such a company. Some people think startups are for crazy people, but they have to be formed by ordinary people [like the ones that work in pharmaceutical firms]. In Japan, we call small firms *Chu-sho-Kigyo* [a small to midsized company; the term connotes being cash-poor and un-

> stable], but in America there's no such word, though they may call it a small company. So now Takeda Pharmaceuticals is going through drastic changes and has purged a lot of people. That was good. The firm got Americanized, and the flip side is that good people got out of the firm, and some of these people started their own firms. . . . So, there are movements, although large firms always seem to block these moves.

According to this professor, the reason why there are fewer startups in Japan is that large firms keep talented individuals, and this problem is exacerbated by biases against small firms. In making this point, he distinguishes between Japanese and U.S. employment practices and industry structures, citing the stable employment practices and economic and cultural dominance of large firms in Japan.

Directly after the last excerpt, this professor drew a connection between contemporary employment practices and an earlier era in Japan's history:

> So do you understand why the Meiji Restoration was successful? [. . .] What made the Meiji Restoration was the Tokugawa Shogunate. Look at Yukichi Fukuzawa, Eiichi Shibusawa—they never thought the Shogunate system would collapse. So they remained in *Bakufu* (the Tokugawa Shogunate political regime), tried to reform it, and then the world changed much faster than they thought it would. The smart, competent men of reason were all within the Shogunate system. They all got laid off, lost their government positions, and started to make Japanese industry. [. . .] So startups will not succeed in Japan unless these talented and sensible people in large firms start to do it.

In this excerpt, he draws a parallel between Takeda Pharmaceuticals' recent restructuring measures and the Meiji Restoration, in which the feudal Tokugawa Shogunate was overthrown to establish a centralized government that could realize rapid modernization and defend against foreign threats. There is a similarity, he implies, between the current situation for Japanese biotechnology startups and a period of Japan's history characterized by rapid societal change and economic development—namely, out-of-date establishments are occupying talented individuals who could otherwise initiate innovation. And this similarity suggests that the current absence of an institutional environment for startups is a fundamentally Japanese problem.

In interviews, Japanese university professors used these psychological, cultural, institutional, and historical accounts to create a myth about the Japanese-ness of university research commercialization in Japan. All the professors I spoke with used institutional accounts to explain the differences between commercialization practices in the United States and Japan. Many also used psychological, cultural, or historical accounts to claim that the way Japanese academia works is fundamentally Japanese, that differences from the American system are necessary and inevitable, and that these differences stem from and prove the fundamental specialness of Japan.

As these examples show, professors took different approaches to nationalizing their explanations of differences between Japanese and U.S. commercialization practices. These examples also show how professors could intertwine and conflate different arguments by connecting mostly accurate explanations of institutional differences to claims about long-standing Japanese cultural traits, ethno-psychological claims about Japanese character traits, and claims that the contemporary situation aligns with Japan's history. It is this upshifting—from institutional settings to culture and essential national qualities—that shaped the ubiquity of nationalizing accounts.[13] Quite simply, I learned that I could expect a nationalizing account along the lines of "You know, this is *Japanese*" at any point in an interview. In this, these scientists resembled the Japanese roboticists who claimed that their social robots were quintessentially Japanese and the medical doctors whose medical practice included asking diabetes patients to stick to "the traditional Japanese cuisine."[14] These scientists, who were enmeshed and successful in the internationalized field of science, perhaps counterintuitively produced very nationalized accounts of why their actions are Japanese.

Settling Down with the New Practices

Japan's adoption of the American research commercialization structure was *made Japanese* through the narratives that scientists and administrators used to nationalize the new practices. These narratives reinforced the idea that "what 'we' do 'here' is different from what 'they' do 'there.'" This distinction, based on national lines, was essential to how Japanese university scientists constructed and maintained their Japanese identities, legitimated their actions, and fostered their sense of belonging to Japanese academia. The narrative of Japanese-ness conflated different claims and concepts. For example, it made no distinction between scientists in Japan and scientists who identify as ethnically Japanese or between institutions created during Japan's modernization period and "essential Japanese-ness." These conflations were fundamental in creating the sentiment that Japanese university scientists operated as *Japanese.* Any differences from what scientists in the U.S. would typically do were attributed to the ethno-national specificity of an exclusive community of the Japanese. The strength of this narrative shows that in what we may think as the most transparently global field—the field of science—everyday nationhood thrives, giving people a sense of self, guiding their social interactions, and shaping their choices.[15]

We should not think of this claim to Japanese-ness as a reaction to the integration and acceptance of the more formalized, U.S.-inspired method of commercializing inventions. As this chapter has shown, the claim to Japanese specificity was an essential part of how Japanese university scientists understood the new practices, modified them, and incorporated them into their toolkits. In saying that what they were doing was Japanese, scientists were claiming ownership of the new ways that Japanese academia was organizing the commercialization of research.

Globalization thrusts a global perspective onto people. Because the adoption of a global policy is always predicated on global competition, "local" actors become very aware of what local practices were before the policy adoption. When enacting new policies, they then become aware of any deviation from the original model. In having to adopt and modify the U.S.-originated rules and practices, comparisons between the United States and Japan became salient; practices that were common in Japan and uncommon in the U.S. became "Japanese." As scholars of translation note, any imported scheme has to be interpreted and made into something that local participants can work with.[16] Everyday nationalism is one way Japanese university scientists understood the new practices as *theirs.* University-firm co-patenting, the juxtaposition of different routes of commercialization, and the continuation of some of the old practices all contributed to scientists' understanding of the new practices as natural, inevitable, and their own.

Ultimately, what the nationalizing accounts did was exactly that—claim and reproduce a compelling sense of nation. It is not that scientific practices in Japan were so unique that scientists had to come up with accounts to explain this uniqueness; it's the opposite: their everyday accounts were what made these practices essentially and naturally Japanese. Japanese deviations from the Western norm are shaped, claimed, and reproduced through daily actions such as accounting for what people do when the rules change. In this particular case of policy imitation, no one I spoke with claimed that the local version was better than the original. Instead, the professors consistently claimed that the Japanese version was *different*—perhaps morally superior in some people's minds, but mostly just of a different nature that specifically catered to the Japanese people. The narrative of Japanese difference thus sustained the circular identity statement that things are as they are in Japan because Japan is *Japanese*.

SEVEN

Conclusion

Variation as Institutionalization

NOW A CLASSIC, Eleanor Westney's 1987 book *Imitation and Innovation: The Transfer of Western Organizational Patterns to Meiji Japan* delineates how Japan's Meiji government so quickly established modern institutions such as a national police system, national postal system, and Western-style newspapers in the wake of the Meiji Restoration in 1868 and the subsequent decision to modernize the country. Japan was an isolated, feudalistic country that had little interaction with the rest of the world before 1859 when it was forced to open treaty ports to Western powers. How did the Japanese government achieve such drastic institutional change and organizational innovation over the next few decades?

Westney found that imitating organizational structures was anything but an easy, cut-and-paste process. In fact, the art of imitation *was* the innovation. Although Japan created its police system, postal service, and newspapers by importing the organizational structures of these institutions from the West, none was a direct reproduction of the Western model, nor could they be. Some parts of these organizations were direct imitations, some were patchworked in from other Western organizational models, some were drawn from the already modernized parts of the Meiji government, and others were adopted from pre-Meiji institutions. As new institutions formed, institutional actors not only retooled traditions to meet new organizational needs but also integrated into the new organizations some "traditions," such as gendered social roles, that the new model didn't possess. The resulting organizational forms were recognizably similar to the modern police systems, postal services, and mass newspapers of the West, yet all of them were different from their original models in substantive ways—for example, the tiny police box would be staffed by only one policeman, because neighborhood safety measures were adapted from Japan's premodern neighborhood policing system, and the post office only allowed men (the legal heads of households) to open accounts. These modern institutions, modeled on the West but also distinct, became taken-for-granted aspects of modernized Japanese society.

Today, Japan and the U.S. are much more similar than they were in the nineteenth century, but the process of importing new structures remains complex. This book's central goal was to explicate the roles Japanese university bioscientists played in establishing Japan's new commercialization structure. Japan's adoption of the U.S.-originated structure of university research commercialization required Japanese university scientists to creatively define and redefine what the new rules would mean in practice. They transformed their practices under the new form without completely altering their practices, relationships, or identities.

As this book has shown, Japanese university scientists creatively altered the new rules as they interpreted and implemented them, including by juxtaposing new practices with older Japanese practices as well as practices they had learned abroad. Many who had worked with firms using the previous gift-exchange-like system felt positively about their experiences and defended the "good old way" as a part of an idealized and long-gone way of doing science that relied upon open communication, scientific curiosity, and a lack of monetary interest on the scientists' side and upon goodwill and donations on the firms' side. The younger scientists I spoke with didn't have direct experience with the donation-based system, but even they seemed to conduct their commercialization activities with attention to this element of Japanese academia's cultural and institutional history. Thus, scientists *negotiated* between the previous gift-exchange-like practices and the new rules to retain some elements of the donation-based systems—such as flexible terms and firms' grip on IP rights—while transitioning to the newer, more formalized practices.

Departing from the conventional wisdom that new policy recipients will resist change, this book has highlighted how creative and cosmopolitan these professional policy recipients—scientists—were. Many of the Japanese university bioscientists I interviewed were *institutional travelers*—actors who had inhabited more than one institution during their careers as professionals in an international organizational field. Japanese university scientists who had worked in U.S. or European academic institutions had cosmopolitan orientations and devised commercialization strategies that included continuing to use the old donation-based practices, pulling the new rules toward established arrangements, innovating practices within the new rules, and engaging in entrepreneurship outside Japan. As my analysis has shown, this juxtaposition of elements from different institutions is itself an innovation.

As a global policy spreads around the world and the new structure settles in different localities, the settling process necessarily includes how the actors affected by the global policy adoption reshape their sense of who they are. When Japanese university bioscientists discussed the Japanese Bayh-Dole Act and subsequent interventions and changes in practices, they described them as essentially Japanese. They engaged in *nationalizing accounts*: they claimed that any and all deviations from the original U.S. rules had derived from Japanese history, culture, and traditions. In their discourse, they produced and reproduced an understanding of the new practices of research commercialization as inescapably Japanese. In doing so, Japanese university bioscientists took ownership of the new practices while successfully maintaining their identity as Japanese academics who lived and worked in Japan.

Inspired by inhabited institutionalism, this book delineated three processes through which the imported structure became institutionalized in Japan: scientists *negotiated* between the previous, gift-exchange-like practices and the new rules; *institutional travelers* used the new rules to strategize solutions to the problems different national contexts presented; and professors used *nationalizing accounts* to incorporate the new practices into a collective identity tied to national-institutional boundaries. The global rules were, then, successfully "naturalized"—modified, institutionalized, and ultimately made "Japanese."

In chapter 1, I asked whether the new American-style policies have replaced the old-style collaboration and whether we should understand Dr. Honjo as the harbinger of the new approach. Although the resulting new commercialization structures and practices look very different from how research commercialization was managed before the policy change, they also don't look like the American structure that was formally adopted. Gift-exchange-like collaboration practices continued, albeit less prominently than they used to.

New formal collaborations between universities and firms took root, but IP arrangements tend to assign more rights to firms than is typical in the U.S. Scientists seem to have no doubt that their professional lives are "Japanese," despite the changes caused by the adoption of the U.S. form and even though some of them practice entrepreneurship outside Japan. Formal continuity is evident, but as identifiable threads and patterns in a new weave. The formal structure is similar to and compatible with the global norm but differs from it significantly in the rules, their operation, and the way people think about the norm.

In the beginning of the book, I drew out two features of Dr. Honjo's press conference after it was announced that he had won the Nobel Prize. First, Dr. Honjo emphasized that the discovery of PD-1 and the following lucrative medical applications of it were grounded in fundamental research and expressed his wish for continued funding of fundamental research. Second, he expressed dissatisfaction with Ono Pharmaceutical, the firm that commercialized his research, for failing to compensate Kyoto University.

After many years of dispute between Dr. Honjo and Ono—and media coverage of the dispute—the two parties agreed to a $230 million settlement to be paid by Ono to Kyoto University's Honjo Memorial Research Fund. At first, the president of Ono seemed rather nonchalant about Dr. Honjo's complaints; he simply said, "We have paid all royalties based on contractual obligations."[1] But pressure from shareholders and other stakeholders eventually led the firm to settle. After the settlement, Dr. Honjo remarked, "I am glad to have established a positive feedback loop where academic contributions and commercialization are valued correctly and help develop the next generation [of] scientists."[2] It would be easy to conclude from this outcome that the new structure of university research commercialization had finally become the norm. In this view, the idea that Ono should pay Kyoto University represents the triumph of the Japanese Bayh-Dole Act; a payment of millions of dollars to a university in exchange for its innovation would seem to have become the new normal.

Part of this interpretation is accurate: the commercialization of research had, by then, become not only a formal part of university life but also a legitimate one. But the economization of innovation is only part of the story. To begin with, throughout the dispute, Dr. Honjo never once mentioned fair compensation for *himself*. Instead, he presented himself as going to all the trouble of suing a pharmaceutical company because he wanted the revenue from his invention to circle back to Kyoto University to enable continued research by younger generations of scientists. In this context, money represents more scientific opportunities, not revenue. In addition, the reason Ono agreed to settle with Dr. Honjo was reputational; stakeholders didn't want the firm to be associated with a fight against the Nobel Laureate.

In other words, the introduction of the U.S.-originated structure of commercialization *did* economize university research in Japan, just as the original policies had done in the U.S.[3] University-firm collaboration became legitimate, scientists now consider it part of university life, and the price of the innovation has to be paid to the university and the university professor. But the policy did not make the exchange between scientific findings and money straightforward. Since the cultural turn, economic sociology has departed from thinking of money as a transparent medium that is clearly distinguishable from other types of exchange. As Georg Simmel pointed out over a century ago, money is more impersonal, less history-laden, and more commensurable to things of value than other mediums of exchange.[4] Putting a price tag on a university scientist's innovation certainly is economizing the research activity compared to, say, gracefully giving away IP rights and pretending to completely forget about it. But still, as Zelizer has pointed out, relational work remains: to establish an exchange between university research and money, the price of the collaboration and the innovation must make sense to all parties, and the exchange *rate* also must make sense to them.[5] There is no simple equation to calculate how much a firm should pay a university scientist who contributed to their product and what that contribution means to the relationships between the firm, the university, and the scientist. At one time, Japanese academia tried to at least pretend that research and innovation had nothing to do with money. Now that the role of money has been acknowledged, financial exchange doesn't simply reflect the unknown market value of an innovation; it also has a *moral* character. When a firm pays for a university researcher's invention, what is the morally right amount, and should the payment be used to fund future researchers, to buy a personal yacht, or to fund a new university building?

To complicate the nature of exchange even further, exchanges often come as a package, or what Gabriel Rossman calls "bundles."[6] In the case of Japanese bioscience, the exchange is not merely financial compensation to a university in exchange for IP rights to its faculty's inventions. These two goods are bundled together with others that may include research collaboration, investment of time, academic accomplishments, public perception and reputation, emotional and personal rewards, risks, financial compensation to the lab, and financial compensation to the scientist.

Japanese university bioscientists are not simply facilitating an economic exchange. They are navigating a new, more commercialized academic environment that involves multiple entangled exchanges and meanings, and that is anything but straightforward. Their own creativity and judgment make the process happen. As new institutional theory recognizes, institutional actors create new practices under new institutional constraints based on their own thoughts and feelings. And these practices have the effect of taming the economic purpose of commercialization by associating it with meanings, such as legitimacy, accomplishment, helping the university, and helping patients. For example, like Dr. Honjo, none of the other scientists I interviewed talked about compensating themselves when they talked about commercialization. Instead, they focused their discussions of commercialization on curing patients, cultivating a more formalized relationship between universities and industry, or ensuring that the university receives a fair share of the profits. What's more, as universities began to emphasize commercialization and industry relations, commercialization and patents became tokens of professional legitimacy. The scientists I interviewed spoke of having to list their patents in their CVs and grant applications and feeling that it was necessary to nod to this new requirement to advance their careers. So while these scientists may have engaged with commercialization more than they did a couple of decades ago thanks to the new environment, it was not only because they wanted to commercialize their research, let alone make money from it. It was also because commercialization became expected of them.

The global rise of commercialization and academic entrepreneurship has to be understood in the context of the complex interplay among global policy diffusion, institutions, cultural shifts, "local" actors (including institutional travelers), and the local environment and its resources. As new institutional theory would predict, the global policy structure and corresponding culture that formalize and encourage interaction between universities and industry has, with some variations, diffused around the world to countries with strong science and technology sectors. But a key theoretical lesson from this book is that the variations in this diffusion are not the result of differences between the formal rules and their implementation, as new institutionalism conventionally conceptualizes decoupling. Instead, the creation of variations is an essential part of the implementation and institutionalization process. What becomes taken for granted in each local space is the particular variation that people in institutions work out while figuring out how to incorporate the new rules into their daily work lives. Institutional actors must not only create a workable version of the new scheme in the new environment but also find new ways of making things meaningful. That is, the new scheme has to not only make sense to them but also align with their professional projects and identity.[7] Japanese university scientists enacted institutional transformation by creating new practices and making them their own by claiming that these practices were Japanese.

The Japanese university bioscientists depicted throughout the previous chapters were by no means passive recipients of global policy diffusion. Scientists' jobs include many obligations. Navigating the ever-changing landscape for doing science is an everyday part of their work lives: they have to keep up with new scientific discoveries and practices, procedures and requirements for grants, the needs of their students, the procedures of their university administration, the norms of university culture, and national and university rules and regulations. When the Japanese government decided to adopt and implement a new set of policies governing research commercialization, that was yet another change that scientists were tasked with handling in their busy work lives. As these scientists created solutions to the dilemmas they faced, they established new practices for commercializing research, and as they repeated those practices, those practices became institutionalized—creating an instance of the local accomplishment of global diffusion. The variation-making, then, was part of institutionalization; the newly developed local rules in Japan are likely to endure even after the scientists who remember the "good old times," because these practices of commercialization will be a new go-to, providing a legitimate template to follow.

The Elusive Institutional Actor, Institutional Theory, and Global Policies

At the book's outset, I explained its theoretical framework: inhabited institutionalism. Looking at Japanese research commercialization policies through the lens of inhabited institutionalism has enabled us to see how Japanese university scientists created a "new normal"—new practices of commercialization that ultimately became taken for granted. They created a variety of new practices including, to name just a few, partially following the rules, university-firm co-patenting, combining the previous and new ways of collaborating with firms, and, in some cases, becoming a science advisor for a U.S. firm. This analysis offers a corrective to new institutional theory.

As I mentioned in the first chapter, new institutional theory's Achilles' heel has been the problem of agency: how can a theory focused on the power of institutional pressures and cultural conformity account for creative action? Anthony Giddens's structuration theory addresses a similar conundrum at the macro level.[8] How, he asked, can we theorize a person's thoughts and actions not as predetermined but as generative and creative when we also claim that the world is governed by rules and resource constraints? How can we explain social change while theorizing a social system? In new institutional theory, this problem trickles down to the question of how institutional changes are possible in organizations. If an institutional actor is simply a carrier of the social grammars of the institution, new institutional theory can only account for changes coerced by the external environment. But since we all know that institutions do change, sometimes with external pressures and sometimes without, how can new institutional theory account for actions that are new, not based on the institutional rules, and original?

The last two decades have been extremely fruitful in building new institutional theories to answer this question. To begin with, Patricia H. Thornton, William Ocasio, and others developed the *institutional logics perspective*.[9] Thornton and coauthors define institutional logics as "the set of material practices and symbolic systems including assumptions, values, and beliefs by which individuals and organizations provide meaning to their daily activity, organize time and space, and reproduce their lives and experiences."[10] According to this theory, it is the plurality of institutional logics in an organization that enables institutional change, as one institutional logic can become more dominant than others. For example, a publisher may have both "market logic" and "editorial logic," but market logic may become increasingly dominant in response to market changes and personnel selection.[11]

In the 2000s, the closely related *institutional work perspective* emerged as an attempt to bring individual actors back into institutional theory. Spearheaded by Laurence, Suddaby, and others, the institutional work perspective highlights the purposive actions that organizational participants take to create, maintain, or disrupt institutions.[12] This perspective aims to account for the effects that individual actors' "idiosyncratic personal interests and agendas for institutional change or preservation" have on institutions.[13] Drawing on Giddens, and others who have furthered theories of agency,[14] the institutional work perspective depicts individuals in organizations as aware of their institutional constraints and able to try to reshape institutions by hypothesizing; that is, individuals creatively try to "reconfigure received schemas by generating alternative possible responses to the problematic situations."[15] Institutional work scholars thus explore the possibility of purposive action by institutionally embedded actors and ask how such actions may reshape organizations.

Lastly, studies of *institutional entrepreneurship* further emphasize actors' ability to initiate and participate in the implementation of changes to existing institutions. Whereas the institutional work perspective was developed specifically as a theory of individual agency, the institutional entrepreneurship literature depicts both individual and organizational actors as continuously and consciously working to alter institutions.[16] Thus, studies of institutional entrepreneurship look for the conditions presented by the organizational field and actors' structural positions that allow actors to innovate.

As I discussed in chapter 1, what distinguishes inhabited institutionalism from other approaches is its theoretical indebtedness to sociological pragmatism and interactionism. In inhabited institutionalism, actors may deliberately and intentionally act, but they are first and foremost theorized as *interacting* and as living their organizational lives. In the course of doing their everyday work, actors negotiate and respond to new situations, and these actions are often largely spontaneous, though they may be just as consequential as strategic actions. Empirically, there are many cases in which institutional changes happen without intentional actors. In the case of a rape crisis center in Israel, for example, the center's gradual institutional transformation from a feminist institution to a more therapeutic one was shaped through managers' pragmatic decisions, volunteers' education and orientation, and the larger cultural and legal environment, which valued professional psychology more than feminist causes.[17] In another case, institutional change toward accountability in a school failed because enforcement of the rules created turmoil by violating teachers' sense of identity and making everyday interactions contentious.[18]

Inhabited institutionalism casts a broad net to capture the microprocesses of institutional change, precisely because so many institutional changes are created through people's situational problem solving. The Japanese university scientists I interviewed are convincing examples of the kind of actors inhabited institutionalism depicts. First and foremost, they were scientists who wanted to do science and, given the opportunity, commercialize their findings. The institution they worked in—the government-regulated university system in Japan—was disrupted by the introduction of U.S.-originated commercialization policies. This disruption presented them with both the need to create new processes and ample opportunities to do so. They created these processes by interpreting their new situation, negotiating its terms, and trying out new ways of doing science. As they devised new practices, these practices became routines, and as the practices became routines, the new institution of university research commercialization became solidified.

They responded to the new commercialization rules without much ado but with many alterations from both what they used to do and what they were supposed to do according to the rules. At first glance, some of their responses may seem to represent different degrees of adjustment to the new rules. For example, "pulling" the new rules to the old practices may appear as an effort to resist the new rules and retain the old practices. By contrast, juxtaposing and variously drawing on the old practices, the new one, and the cosmopolitan ones they learned in other countries may appear as a complete adjustment as scientists nimbly transcend the boundaries where any particular set of rules applies. But both responses are scientists' pragmatic solutions to the situation they find themselves in; they are simply different configurations of the same loose coupling of structure and practices, and that explains why the same university scientist could use the informal practices in one occasion and get involved in a U.S. startup in another commercialization effort. This loose coupling is enabled by the scientists' multiple experiences and resources, including their experiences with different rules and institutions, and by an organization—the university—that gives them room to experiment.

In explaining how scientists' international experiences affected their responses to change, I introduced the concept of *institutional travelers.* This concept advances the explanatory value of inhabited institutionalism by connecting individuals' trajectories and social interactions. Scholars who focus on institutional entrepreneurship and institutional work tend to treat individuals as the primary, and intentional, agents of institutional change. By contrast, scholars who draw on inhabited institutionalism tend to focus on the interactions that take place in organizations and downplay individuals themselves. But social interactions are shaped by individuals, who come to the organization with their own backgrounds, trajectories, and lives outside the organization, all of which affect how they interact in the organization. Because institutional travelers gain experience in different institutions over the course of their careers, they can play an important, if not predictable, role in understanding, adapting, and modifying a new imported structure and making it local.

The case of research commercialization in Japan cannot be fully explained as the result of either scientists' purposive actions or the dominance of an economic logic, as the other approaches to institutional change would have suggested. The institutional entrepreneurship and institutional work approaches would try to identify the change agent, but the new practices that emerged weren't a product of anyone's promotion or coercion. We could explain the case as economic logic becoming dominant in Japanese academia, as the institutional logic perspective would do. But while this explanation is certainly true to some extent, it does not help us understand how this logic's dominance was accomplished or how the actors who had to deal with this new situation created a new set of practices that works for them. In other words, while these approaches are helpful, they don't fully explain the processes through which the institutional transformation happens and how institutional actors shape that transformation.

Having reclaimed the agency of institutional actors, the next intellectual project of the "new" new institutional theory—including inhabited institutionalism—is to reintegrate macrotheories of global diffusion, such as the world society thesis, which co-evolved with new institutional theory.[19] In regard to global policy adoption, this book offers several insights into how global structures are recursively made and remade as they diffuse to country after country. To understand this process, it's necessary to attend, first, to the dynamic between the original policy and what happens in the adoptee countries and, second, to the relationship between global professions and global policy. The U.S. policy of research commercialization has diffused worldwide. And while this diffusion implies a seeming uniformity, there is clear evidence that not a single country that adopted this policy became a copy of the United States by any relevant metric, including the number of startups with university-related inventions, the way IP rights are managed, the number and scale of university-firm collaborations, the number of inventions coming out of the universities, and so on.[20] Rather than thinking of this variation as indicating a failure of the global form, this book's analysis guides us to understand that the creation of variations *was* the institutionalization; that is, local actors' meaning-making and reshaping of the form were necessary conditions for the global policy's successful implementation. Such local interpretations and variations are influential; they can strengthen the legitimacy of the original form[21] and in some cases even alter the original rules.[22]

This book also highlighted one additional kind of interplay between a diffusing global policy and local adoption and adaptations: global professionals who move between institutions and organizations across countries. Academics, in this case, occupy a specific social position as cosmopolitans within the global field of science. These transnational actors can be the harbingers of new schemes and the elites who make local policies upon their return.[23] They can also be the recipients of the new policy, as they are in this book. Institutional travelers embody knowledge about and ties to the place where the global form originally started and has been practiced. These actors can apply elements of the new forms and creatively juxtapose different elements, further contributing to the unique variations that the global policy adoption introduces. Going back to Westney, skillful juxtaposition is an act of innovating.[24]

Such reintegration of global diffusion of forms and new institutional theory that takes actors seriously enriches our understanding of the conditions under which certain types of variations emerge and how and whether these variations will be successfully institutionalized in the local space. For example, Kiyoteru Tsutsui delineates how resident Koreans in Japan expanded their civil rights after the country's ratification of the International Covenant on Civil and Political Rights and the International Covenant on Economic, Social, and Cultural Rights. Here I highlight the three factors that led to the institutionalization of noncitizen rights through the analytical lens of inhabited institutionalism developed in this book: the actors' experience with activism prior to the ratification; the actors' international exposure, including traveling to the U.S. through Korean Christian organizations; and their successful appeal to the Japanese public by arguing that Japan needed to uphold human rights to be an honorable member of the international community.[25]

The strength of the UN's human rights advocacy depended in part on a "weak" form: underspecified, norm-based rules from a high-legitimacy organization that allowed local innovation.[26] In this book's case, while the form—the set of legal and administrative rules for commercialization—was specific, its local adoption necessitated modification due to the different environment surrounding commercialization, and the loose coupling between rules and practices was implicitly accepted. In both cases, the institutionalization of the variation was successful—the variation became taken for granted. Further research is needed to theorize the relationship between the global form's characteristics, the local institutional environment, and the recipients' ability and propensity to create a successful variation that can be institutionalized.

On a more micro level, this theoretical move to reintegrate new institutional theory and globalization allows us to see how institutional actors are interacting with the new rules and with other actors in changing situations. It additionally allows us to appreciate the recursively global nature of policy adoption and professional movement. Institutional actors learn and embody repertoires of actions from different institutions internationally, and when a global form is adopted, they innovate. They partially neglect, modify, transpose, and patch together different elements of the previous system, the new system, and the systems from elsewhere to create new practices that are workable enough to become taken for granted in their local space as time passes and successful practices accumulate.

APPENDIX A Confessional Tale of Theory, Methods, and Positionality

A METHODOLOGY APPENDIX PROVIDES a sneak peek into an intellectual adventure and the decisions the researcher made through the process—how they entered into a different social world such as an immigrant community, a grassroots political group, a hospital, or a poor neighborhood. Less heroically, in this appendix I offer my own account of the circumstances that led me to collect the interviews I use in the book, how I analyzed them, and my position in relation to the interviewees who kindly agreed to meet with me.

My fascination with ethnography and qualitative methods started with a lecture by Professor Toshiaki Kimae at Osaka University when I was an undergraduate studying sociology. Reading Paul Willis's *Learning to Labour*, I was amazed by the depth and richness of the author's analysis and fascinated by ethnography as a method that can delve into people's everyday lives and their ways of experiencing the world.[1] My graduate training at UCLA, where Chicago-school interactionist ethnographers mingled with ethnomethodologists and conversation analysts, further prepared me to try to see the world as my study participants saw it. I imagined myself becoming part of the new generation Chicago-school-style ethnographers, gaining access to unexplored places and hanging out with distant "others."

This was not, however, how things worked out. I didn't end up conducting ethnographic work, nor did I study distant others: I studied academics in universities and the relationship between university science and industry in Japan. More specifically, I studied how Japanese university scientists reshaped their practices after the rise of university research commercialization. I traveled back to Japan; I was very much "close to home."

When I developed the idea for the research detailed in this book, I was working as a research assistant for Prof. Lynne Zucker on the role of universities in innovation. I initially intended to do an ethnography of a university lab—I believed (and still do) that all good interactionists should conduct ethnographic fieldwork. Yet ethnographic work is not always the best way to get at the questions we care about. Commercialization of research is only a small part of a scientist's daily routine, and it would have been impractical to spend a week in a lab in the hopes of observing an occasional twenty-minute conversation the professor may (or may not) have with an industry partner. In other words, practically, some events that are essential to an organization are not easily observable because they happen infrequently, over long stretches of time, behind closed doors, or in some combination of the three. Especially for an inquiry on long-term *trajectories*, in-depth interview design was the best method. As one of my advisors at the time, Stefan Timmermans, succinctly put it, "You can't do ethnography, but you can interview them."

Still, my approach to my research, and to the interview design, was guided by interactionist ethnographic methodologies, from Barney G. Glaser and Anselm L. Strauss's *The Discovery of Grounded Theory* through the work of interactionist methodologists such as Stefan Timmermans, Bob Emerson, and Jack Katz.[2] In both reading and interacting with these methodologists, I learned the value of asking "how" instead of "why." In one memorable methods session, we asked each other questions, formulating them in alternate ways—both as "how questions" and as "why questions." The "how" question invariably elicited so much more about the processes through which one situation propelled the interviewee to another situation, where they then made new decisions or encountered a new set of actors. I also learned the art of in-depth interviewing—one would have an interview guide with a list of questions, but the interview should never go from the first question to the last question, nor be confined to what was on the page; when interviewees want to talk on, seemingly even for irrelevant details, you must allow it to take its own winding path.[3] And then, there were books that influenced the way I thought about interviewing and what one can do with interview studies. *Talk of Love*, by Ann Swidler, showed me how we can use interviews to understand how people mobilize different cultural frameworks to evaluate situations and respond to them.[4] *Uncoupling*, by Diane Vaughan, dealt with the processes of couples separating using in-depth interviews. Interviewing about *trajectories*—how one action led to another, a response to a situation at a time—was exactly what I wanted to do, and these books' interpretive orientation helped me in structuring my own book.[5] Although the two books' substantive interest—love and separation—was completely different from my own, they provided exemplars, showing how to do interactionist work using interviews and what the results of the analysis could look like.

The book is primarily based on in-depth interviews conducted between 2010 and 2017 as well as archival research. The archival materials included official policy notifications and other documents from the Ministry of Education, Culture, Sports, Science, and Technology (referred to as MEXT); the Ministry of International Trade and Industry (MITI); the Ministry of Health, Labor, and Welfare (MHLW); and their related semigovernmental organizations (Japan Science and Technology Agency, New Energy and Industrial Technology Development Organization, and Pharmaceuticals and Medical Devices Agency, Japan). Additionally, I collected university administrative, and private sector reports on commercialization and technology transfer and media coverage of university research commercialization.

While the archival materials both contextualized my research and suggested some questions I hadn't thought about—for example, the legal implications of university-firm co-patents were something that I only really grasped as I read through the documents—the bulk of the analysis comes from interviews with sixty-three people, including fifty-five prominent Japanese university bioscientists, five university administrators involved in university-firm collaboration or technology licensing, and three firm scientists who had worked closely with university scientists. Some of the scientists interviewed also held governmental or university administrative positions—for example, as board members of the Council of Science and Technology Policy—in which case I also asked them about their work in that domain. I also informally met with a few governmental agency administrators.

I selected university scientists who are highly successful and influential. This was a theoretical decision. Commercialization policies target those high-profile scientists because commercialization is most successful when "star scientists" from academia are involved in the process.[6] Moreover, studies show that actors and institutions with strong reputations influence others to accept new practices and institutional logics.[7] In other words, I targeted scientists in top research universities.

I then used several criteria for selection to diversify the institutions and the subdisciplines in which I conducted interviews as well as acquire enough interviews with those professors who were often extremely busy. First, I contacted *ISI Highly Cited* scientists whose home institution was in Japan. Secondly, I looked for influential scientists who worked at one of *Times Education*'s two hundred World Top Universities between 1990 and 2010 and who were also engaged in commercialization activities. I also made an effort to interview scientists from different geographic areas, interviewing scientists in the broader Tokyo Metropolitan Area, Tsukuba (a research university town near Tokyo), the Kansai area, and Nagoya area. Interviewees were then selected by convenience sampling, and the response rate was about 60%. Their level of commercial involvement varied from having a co-publication with a firm (signaling that they interact with firms) to having their own startups. Additionally, I interviewed policymakers, administrators of universities' Offices of Research or university technology transfer office personnel, and firm scientists. To maintain confidentiality, I only mention interviewees' positions and general areas of research throughout the book.

I implemented an ethnographic interview method, consisting of both *shared questions* and *individually tailored questions*.[8] I had an interview guide that asked the interviewees to recall their collaboration with industry from the beginning of their career—if possible, from their PhD training—and asked them about their patenting activities. Following through their trajectories as an academic required an interviewee to specify the actual processes of their work and the history of their involvement with commercial firms and/or their own entrepreneurial activity as their career has proceeded in the context of shifting national policies, university regulations, and departmental cultures.

Yet in addition to asking them to narrate their career in commercialization in loose chronological order, I also did extensive research on what area of biosciences they were in; what specific topics they had been working on; which patents they tried to acquire, with whom, and if they succeeded or not; and their more time-consuming commercial involvements, such as being a scientific advisor or on a board of a startup. Japanese university bioscientists do not post their CVs online, but there was ample public information for a scientist of their status. I often used the university's website, where the lab often had its own website detailing the professor's accomplishments, various scientific publication search engines such as PubMed or Google Scholar, Japan Patent Office patent search, United States Patent and Trademark Office (USPTO) patent search, and searches by major internet browsers. The USPTO patent search was useful for two reasons: international patenting signals serious intent of commercialization, and it compensates for the coverage of the Japan Patent Office patent search, which only had partial coverage for older patents.

Because many of the scientists had more than thirty years of career behind them, it was helpful to have this information to help jog their memories and structure the interview. This combination of the interview guide and individually tailored research of the scientists' work was inspired by Harriet Zuckerman's (1977) classic work in the sociology of science, *Scientific Elite: Nobel Laureates in the United States*. In her book, she described the thorough research she conducted on the scientists' work before interviewing them and how that helped her gain credibility to converse with the scientists immensely. Following Zuckerman's lead, I spent a few days poring over each scientist's research and their discovery that was the basis of their commercialization efforts—trying to at least reach the level of science journalism for the public, something akin to an educated layperson's effort to understand how mRNA vaccines work in the human body postpandemic. The interviews usually lasted between 30 and 180 minutes, most of them falling between 45 and 75 minutes. All interviews were conducted in Japanese, and the quotes in this book are translations by the author.

I collected the majority of interviews over the summers, usually in July or August. As it turns out, Japanese academics are almost invariably present on campus during these months, except for the brief *Obon*—a few days of official summer holidays where custom dictates one honor the ancestral spirits by taking a short vacation. During these months, walking down the spacious campus to whatever building—the biotechnology building, agriculture department, medical school—I was drenched in sweat, embarrassed that I looked desperate. Many times, the secretary of the professor—always a woman—would bring me a cold green tea as soon as I sat down. In some cases, the professor himself—always a man—would pass me a bottle of cold green tea. That all scientists I interviewed were men is in line with gender inequalities in Japan, especially in science and in older generations of scientists.

I am grateful that such busy people had graciously given me time; more than a few of them had been nominated for the Nobel Prize by 2022. My ability to interview the elite Japanese university bioscientists was undoubtedly aided by some biographical features. I was affiliated with a U.S. academic institution, the University of California–Los Angeles (UCLA), which interviewees regarded as prestigious. The scientists seemed to accept media interviews and other inquiries as part of their unwritten responsibilities as professors of a public status, which surely helped, but many were curious about my status as a budding social scientist in the United States. As chapter 5 had shown, many of these professors also recognized that time studying abroad is an inevitable, and exciting pathway to academic success, and they shared their own experiences as something I'd also understand, despite our disciplinary differences. I felt like a temporary and honorary insider during many interviews. My position probably mattered most regarding the scientists' conviction that their commercialization practices were quintessentially Japanese. During data coding, I noticed that it was likely that the scientists took it for granted that I, too, understood Japan as special. As a native Japanese woman, I fit the category of a co-ethnic, co-national listener. Because of the way the talk assumed my understanding of Japanese authenticity, I suspected that I had inadvertently initiated this topic. I thus performed an additional round of data examination to make sure that I did not bring up Japanese-ness or attribute the reason for the scientists' practice to the ethno-national category myself.

I recorded and transcribed all interviews and analyzed all transcripts and documents using a modified version of "grounded theory."[9] I have read the interviews closely multiple times, and went through the stages described in Emerson et al., open-coding, writing analytic memos, and finding themes.[10] Some of the early themes included *entrepreneurship, good old days, comparing to the U.S., gentlemen's agreement, patenting, creating medicine*, and *university as an intruder.* I wrote dozens of analytic memos to organize my thoughts and narrow down the themes of interest, also informed by my expertise in organizational theory, having completed previous research on how software engineers' informal culture transformed what was supposed to be an "empowering scheme" into a source of doubt.[11] After multiple codings and memos, I decided to focus on the themes that are present in the book.

Interviews were full of surprises. In what became a major theme in the book, I didn't expect that the older collaboration practices of working with firms informally, and giving them IP rights as a gift, would be such a large part of scientists' commercialization life. Most literature focused on how the new policies that were implemented from 1999 to 2004 increased university-firm collaborations and university patents, measured by the increase in the number of patents that had a university (or a TLO) as one of the assignees. Yet in the very first interview with a professor of molecular internal medicine, he remarked that "twenty years ago university professors didn't think about being involved in commercialization themselves" and proceeded to tell me how pharmaceutical companies would help him patent, and send some researchers to his lab if they wanted to research more, with *shogaku-kifu* research donations. As I made progress in collecting more data and familiarized myself by reading interview transcripts numerous times, it was increasingly obvious that I had to focus on how the university scientists shaped their new commercialization practices after the policy intervention in relation to gift-exchange-like practices that were both ubiquitous and meaningful.

Grounded theory, theoretically, asks the fieldworkers to be as close as possible to an idealized tabula rasa, an unbiased observer taking indigenous meaning at its word. That, however, is in no way possible. As Emerson et al. observed, because the ethnographer will inevitably filter a lot of things out, it's an *ethnographer's account* of indigenous meaning.[12] Moreover, it is simply unsustainable to argue that one didn't have any preconceived ideas about the case when that same person took sociology courses; has already been writing about economic sociology, organization theory, and ethnography (or whatever their subfield is); wrote detailed interview guides, and so on. To the researcher's relief, recent methodological debates about the potential of data-driven qualitative analysis acknowledge what good ethnographic researchers already knew: the iterative and recursive nature of moving between data and theory. *Abductive analysis*—a rethinking of grounded theory based on American pragmatism, especially that of Charles S. Peirce—"rests on the cultivation of anomalous and surprising empirical findings against a background of multiple existing sociological theories and through systematic methodological analysis."[13] Abductive analysis encourages researchers to be steeped in existing theories during all phases of research, from start to finish. It requires researchers to look for unexpected research findings and first develop speculative theoretical hunches that explain these findings and then advance the emergent theories with a systematic analysis of variation across a study.

In my case, my theoretical background was new institutional theory, especially that of inhabited institutionalism.[14] I recursively moved between data and theory by systematically coding the interview notes and archival materials to find themes to arise inductively, but I also attended to literature on practice change in organizations, science and technology policy, and institutional theory. Some themes that arose fit well into the preexisting theory of new institutional theory, such as the decoupling of the rules and practices. Others were surprises—for example, I did not expect the theme of "Japanese-ness" to arise until after multiple stages of data coding. This did not fit well with new institutional theory, even with inhabited institutionalism that posits that members' negotiations of meaning were an integral part of an organization. After iterative processes of interview data analysis and attending to a broader sociological literature on global policy diffusion and everyday nationalism, I developed my theoretical approach. Inhabited institutionalism gone global, my theoretical framework in this book explains why the previous relationships and practices, the institutional travelers, and nationalizing accounts are the three pillars of successful global policy adoption and adaptions when global policy travels to environments vastly different from its original site.

Notes

Chapter 1

1. Tasuku Honjo, "Nobel seirigaku igakusho jyushoude Honjo Tasuku tokubetsu kyoujyu ga kaiken 1" [Specially appointed Professor Tasuku Honjo holds a meeting after the announcement that he was awarded the Nobel Prize in Physiology or Medicine 1]. ANNnewsCH, accessed November 21, 2022, https://www.youtube.com/watch?v=-3WbonGe1eQ. Transcribed by the author.

2. This date refers to the patent application in the U.S. He applied for a Japanese patent in 1994.

3. Marketed in 2014 under the trade name Opdivo.

4. Tasuku Honjo, interview with Prof. Tasuku Honjo. Throughout the book, bracketed ellipses indicate an omission in the interview quotes.

5. The lawsuit filed by Dr. Honjo was not directly about the contribution that Ono should give to Kyoto University for patents on PD-1. It was about Dr. Honjo's involvement and the allegedly promised compensation for his help in Ono's lawsuit against Merck on patent infringement related to PD-1. Dr. Honjo, however, was vocal about the purpose of this lawsuit for him: making firms properly compensate for university-firm collaboration so future researchers would be properly rewarded. In the end, Ono settled with Dr. Honjo and agreed to voluntarily donate a research fund to Kyoto University, accepting Dr. Honjo's requests.

6. The total revenue through sales of Opdivo in Japan. Ono Yakuhin Kougyou Kabushiki Gaisha 2018.

7. See *Asahi Shimbun* 2018.

8. See Sapir and Kameo 2019.

9. Koyama 2017.

10. One important quantitative aspect, which I will return to in the next chapter, is the number of patents for inventions that involved university scientists. To measure that change, we would need to identify not only patents resulting from formal collaborations but also those born from informal collaborations between university scientists and firms, which do not name universities as assignees.

11. See AUTM 2017.

12. See Mowery et al. 2001.

13. Meyer and Rowan 1977, 341.

14. For research along these lines, see, for example, Boli, Ramirez, and Meyer 1985; and Scott and Meyer 1994.

15. See DiMaggio and Powell 1983; Kelly and Dobbin 1998; and Meyer and Rowan 1977.

16. Meyer and Rowan 1977, 357.

17. For important work along these lines, see, for example, Czarniawska and Sevón 1996; Djelic 2001; Djelic and Sahlin-Anderson 2006; and Drori, Meyer, and Hwang 2006.

18. See Chorev 2012.

19. See Hallett 2010.

20. See Stinchcombe 1997; see also Perrow 1986, 157–218; and Selznick 1996.

21. For the former, see DiMaggio and Powell 1983; and Hirsch and Lounsbury 1997. For the latter, see, for example, Selznick 1949.

22. For critiques along these lines, though coming from slightly different angles, see Bechky 2011; DiMaggio and Powell 1983; Greenwood et al. 2008; and Hirsch and Lounsbury 1997.

23. Powell and Colyvas 2008, 277.

24. See Hallett and Hawbaker 2021; and Hallett and Ventresca 2006a; 2006b.

25. Czarniawska and Sevón 1996.

26. For somewhat different theorizations of the way actors act within organizational settings, see Binder 2007; Hallett 2010; and Zilber 2002.

27. See Hallet and Ventresca 2006b; Friedland and Alford 1991; Greenwood et al. 2011; Lounsbury 2007; and Thornton, Ocasio, and Lounsbury 2012.

28. See Binder 2007; Berman 2008; Haveman and Gualtieri 2017; and Thornton, Ocasio, and Lounsbury 2012.

29. McPherson and Sauder 2013; also see Zilber 2002.

30. See Hallett and Hawbaker 2021.

31. Hallett 2010; Hallett and Hawbaker 2021; Leibel, Hallett, and Bechky 2018.

32. Berman 2012.

33. Merton 1973.

34. See, for example, Etzkowitz 1989; 1998; Nelson 2014; Slaughter and Rhoades 2004; and Johnson 2017.

35. For this aspect of Zelizer's position, see Zelizer 2005; 2011; 2012.

36. See Healy 2005; and Zelizer 2011.

37. Lukes 1992.

38. Toole and Czarnitzki 2007.

39. See Slaughter and Leslie 1999.

40. Evans 2010.

41. $43.8 million and 23 billion yen, approximately, at the exchange rate of the time.

Chapter 2

1. See Office of the Cabinet 2015.

2. Westney 1987.

3. See, for example, Cox 2008; and Cusumano 1988.

4. Chorev 2012; Meyer et al. 1997.

5. See Boxenbaum 2006; Inglehart and Norris 2003; and Stamatov 2013.

6. Djelic 2001; Djelic and Quack 2003; Drori et al. 2003.

7. See Strang and Meyer 1993.

8. See Johnson 2017.

9. See Berman 2008.

10. See Mowery et al. 2001; and Stevens 2004.

11. See Berman 2008; 2012; Colyvas and Powell 2006; Metlay 2006; Mowery and Sampat 2001; and Slaughter and Rhoades 2004.

12. See Berman 2008; Henderson, Jaffe, and Trajtenberg 1998; and Mowery and Sampat 2001; 2002.

13. See Berman 2008.

14. See USPTO 2022.

15. See, for example, Colyvas 2007; Etzkowitz 2002; and Hughes 2001.

16. Cohen et al. 1973.

17. Berman 2012.

18. Berman 2008; 2012.

19. Berman 2008; Mowery et al. 2001.

20. See Colyvas and Powell 2006; Johnson 2017.

21. Britain was particularly swift to transition to institutional ownership of IP rights because of its own biotechnology advances that ended up not patented.

22. For an argument along these lines, see Sampat 2006.

23. See So et al. 2008.

24. Gordon 2002.

25. Ministry of Education 1962.

26. The other slogan was *Shokusan-kogyo*, or "promotion of industry development."

27. Although the number of private universities well exceeds that of national universities, most of them are focused on education, not research. According to MEXT, there are nineteen "Research Universities" (roughly equivalent to R1 Universities in the U.S.), of which only two are private (Keio and Waseda Universities.)

28. See Branscomb, Kodama, and Florida 1999; Kamatani 2006; and MEXT 2021.

29. See Amano 2009.

30. Notification no. 117, Ministry of Education 1978.

31. See Kneller 1999.

32. See Ministry of Education 1964.

33. Yoshihara and Tamai 1999.

34. Ministry of Education 1997, in Odagiri 1999.

35. See Pechter and Kakinuma 1999.

36. See Zucker and Darby 2001.

37. See Kameo 2015.

38. *1997 nen kougyoushoyuuken shinsa gikai songaibaishoutou shouiinkai houkoku* [1997 report by the committee for the examination of industrial property regarding the compensation for damages].

39. See *Asahi Shimbun* 2010.

40. In 2001, the Ministry of Education, Japan, was succeeded by MEXT, an official abbreviation of the Ministry of Education, Culture, Sports, Science, and Technology, Japan. MITI refers to the Ministry of International Trade and Industry, Japan.

41. Collaborative projects (called Kyoudo Kenkyu) and sponsored projects (Jutaku Kenkyu) are two types of contracts between a Japanese university and a firm. Roughly speaking, the difference between the two is based on whether firm scientists are actively collaborating in the research or the university lab is a contractor for conducting specific research.

42. See MEXT 2019.

43. See also MITI 2006; and Yamaguchi 2010.

44. See Yamaguchi 2010.

45. MEXT 2014.

46. Although the streamlined official mechanisms for university patents became available only in the 1990s, there were some mechanisms for university scientists to rely on for university patenting without involving firms before the 1990s. Those include the alumni associations for select departments in a few universities—such as the Shirankai for Kyoto University medical school.

47. *Heisei 15 nendo Daigaku Touni Okeru Sangaku Renkeitou Jisshi joukyou ni tsuite* [The situations on university-firm collaboration, 2003], accessed August 22, 2022, https://www.mext.go.jp/a_menu/shinkou/sangaku/sangakub/04072301/004.htm.

48. Kanama and Okuwada 2008.

49. More specifically, 13%, 20%, and 25% of the patents applied were single-handedly owned by firms in Hiroshima, Tohoku, and Tsukuba Universities, respectively.

50. See Kneller 2011.

51. MEXT 2020.

52. Wimmer and Glick Schiller 2003.

53. MEXT 2020.

54. See Kneller 2011.

55. Kneller 2003.

Chapter 3

1. The name is a pseudonym to ensure confidentiality.

2. See Branscomb, Kodama, and Florida 1999.

3. See Zelizer 1994.

4. For the classic work on "The Gift," see Mauss (1925) 1954.

5. See Zelizer 1994; 2013.

6. See Mauss (1925) 1954; Rossman 2014.

7. See Rossman 2014.

8. Ministry of Education 1984a.

9. Ministry of Education 1984b.

10. Kameo 2015; Kneller 2003.

11. Koyama 2017.

12. See Zelizer 2005.

13. For the notion of "bundling" of exchanges, see Rossman 2014.

14. See Berman 2012.

15. Zelizer 2013.

16. Kneller and Shudo 2008.

17. See Kneller 2003. Unlike in the U.S., the Japanese patent laws require all assignees to agree to license the patent.

Chapter 4

1. Meyer and Rowan 1977.

2. Westney 1987.

3. This arrangement is in fact common in other countries that imitated the Bayh-Dole Act, such as Germany or Italy.

4. See Ledford 2013.

5. The irony here is that professors routinely gave the rights to their inventions to firms before, but it was never officially acknowledged. Officially, inventions arising from faculty research were owned by the faculty member (if the faculty member did not choose to make it public), and research funded by national agencies was either considered national invention or, based on negotiations, patented through national agencies and the Japan Science Technology Agency (JST), a public entity that handled patents arising from nationally funded research before the Japanese Bayh-Dole Act. In other words, inventions were patented, just not through universities.

6. See Branscomb, Kodama, and Florida 1999; and Kneller 2003.

7. Cabinet Office of Japan 2004.

8. MEXT 2017.

9. See Meyer and Rowan 1977.

10. See also Kanama and Okuwada 2008.

11. University of Tokyo 2011.

12. MEXT 2016.

13. See also Berman 2008; 2012; Colyvas 2007; Colyvas and Powell 2006; Etzkowitz 1983; 1989; 1998; Powell and Colyvas 2008.

Chapter 5

1. See Hallet and Ventresca 2006b; and Friedland and Alford 1991. See also McPherson and Sauder 2013. For recent work in institutional logics, see Greenwood et al. 2011; Lounsbury 2007; and Thornton, Ocasio, and Lounsbury 2012.

2. McPherson and Sauder 2013.

3. See Scott 1995; Zuckerman 1977.

4. Such agreements were often formalized by creating a semilegal document called an *oboe-gaki* (memorandum of understanding) between the individual (scientist) and the firm. Universities are not involved in formulating or maintaining such documents. This memorandum was also widely used for agreements of confidentiality in collaborative research.

5. The difference between Japanese and U.S. patent laws makes this arrangement even more advantageous to the firm in Japan. Whereas U.S. patent law allows each patent assignee to make, use, and sell the patented invention without the permission of others, Japanese patent law dictates that all co-assignees of the patent must agree prior to licensing. As a result, the firm essentially has an exclusive and royalty-free right to exercise the patent. This is probably why U.S. universities are extremely reluctant to agree to jointly hold IP rights with firms—the firm may decide to license the IP without the university's consent. One study done by Kneller (2011) that investigates U.S. patents issued between April 2008 and March 2009 shows that among 2249 U.S. patents that had at least one U.S. university as an assignee, only 3% had at least one private corporation as a co-assignee.

6. There are several possible explanations as to why Japanese universities allow arrangements like this one. First, especially around the early 2000s, when the Hiranuma Plan for one thousand university-originated startups was rolling out (2001), universities were eager to establish university spin-offs rather than negotiate for their own pecuniary interest. The need to "meet the numbers"—be it the number of patents or the number of startups—was a recurrent theme that scientists and administrators raised during the interviews. For example, universities were quick to compromise on the details of co-assigned patents because they needed to meet target numbers.

7. For work on such "returnees," see Wang 2014.

8. See Gouldner 1957.

9. See Gouldner 1957; and Merton 1973.

10. See Glaser 1963.

Chapter 6

1. See MITI 2023.

2. See DiMaggio and Powell 1983.

3. See University of Tokyo 2011.

4. For the theoretical roots of the literature on "doing being," see Garfinkel 1967; and West and Zimmerman 1987.

5. See Tisdell 1981.

6. See, for example, Nelson 1993.

7. See Goldstein-Gidoni 2001.

8. See Kreiner and Schultz 1993; and Powell, Koput, and Smith-Doerr 1996.

9. D'este and Perkmann 2011; Owen-Smith and Powell 2001.

10. See Parmentier 1994; and Timmermans and Tavory 2020.

11. See Yoshino 1992.

12. See, for example, Doi 1971.

13. For "upshifting" in a different context, see Timmermans and Tavory 2020.

14. For the notion of Japanese robots, see Sabanovic 2014. For an analysis of the "Japanese" diabetes, see Armstrong-Hough 2018.

15. See Bonikowski 2016, 428.

16. See Czarniawska and Sevón 1996; Djelic 2001; Djelic and Sahlin-Anderson 2006.

Conclusion

1. See *Asahi Shimbun* 2018.

2. "Gan meneki chiryouyaku Obdivo o meguri taijishita Honjoshi to Onoyakuhin saiban o tsuujite tsutaetakatta Honjoshi no jisedaikenkyuusya he no omoi" [The reason for the Prof. Honjo and Ono Pharmaceutical trial regarding the cancer immunotherapy Obdivo: Prof. Honjo wants to encourage the next generation of researchers], MBS news, January 28, 2022, accessed November 22, 2022, https://www.mbs.jp/news/feature/kodawari/article/2022/01/087487.shtml.

3. Berman 2012; Johnson 2017.

4. Simmel (1900) 2011.

5. Zelizer (1979) 2017.

6. Rossman 2014.

7. See Turco 2012.

8. See Giddens 1986.

9. Thornton and Ocasio 1999.

10. Thornton, Ocasio, and Lounsbury 2012, 2.

11. See Thornton, Ocasio, and Lounsbury 2012.

12. See Greenwood, Suddaby, and Hinings 2002; Lawrence, Suddaby, and Leca 2009; and Zilber 2002.

13. Lawrence, Suddaby, and Leca 2009, 6.

14. See Giddens 1979; 1986. See also Emirbayer and Mische 1998.

15. Emirbayer and Mische 1998, 984. See also Battilana and D'Aunno 2009.

16. Battilana, Leca, and Boxenbaum 2009; Fligstein 2001; Greenwood, Suddaby, and Hinings 2002.

17. See Zilber 2002.

18. See Hallett 2010.

19. See Meyer et al. 1997. See also Colyvas and Jonsson 2011; Drori, Meyer, and Hwang 2006; and Lim and Tsutsui 2012.

20. See, for example, Geuna and Rossi 2011; Haeussler and Colyvas 2011; and Sapir and Kameo 2019.

21. Fourcade 2006; Hafner-Burton and Tsutsui 2005.

22. Chorev 2012.

23. Czarniawska and Sevón 1996. See also Fourcade 2006.

24. See Westney 1987.

25. Tsutsui 2018.

26. Tsutsui 2018.

Appendix

1. See Willis 1978.

2. Glaser and Strauss 1967.

3. See Weiss 1995.

4. Swidler 2001.

5. Vaughan 1986.

6. See Zucker, Darby, and Armstrong 2002.

7. See Owen-Smith 2011; and Stuart and Ding 2006.

8. See Spradley 1979; and Weiss 1995.

9. See Glaser and Strauss 1967; Emerson, Fretz, and Shaw 1995; also see Timmermans and Tavory 2007; 2012.

10. See Emerson, Fretz, and Shaw 1995.

11. Kameo 2017.

12. Emerson, Fretz, and Shaw 1995.

13. Timmermans and Tavory 2012, 169; also see Tavory and Timmermans 2014.

14. See Hallett and Hawbaker 2021; Hallett and Ventresca 2006a; 2006b.

References

Amano, Ikuo. 2009. "Nihon no daigaku kaikaku" [Reforms on Japanese universities]. *Koutou Kyouku Janaru* [Journal of Higher Education] 3:58–64.

Asahi Shimbun. 2010. "Gijyutu iten kikan ni akikaze: raisensu syunyu teimei" [Autumn wind blows on technology transfer centers: Licensing revenue stagnates]. September 18.

———. 2018. "Dare mo yaranai bunya ni chousen: Opdivo kaihatsu Ono Yakuhin shacho noberu syou Honjo shi no shiteki ha 'shingai'" [Surprised with Dr. Ono's Remarks: Interview with the President of Ono Pharmaceutical on Opdivo Development—Dr. Honjo's comment was "unexpected"]. Osaka edition. October 27.

AUTM (Association of University Technology Managers). 2017. "FY2017 Licensing Survey." Accessed February 10, 2024. https://autm.net/AUTM/media/SurveyReportsPDF/AUTM_2017_US_Licensing_Survey_no_appendix.pdf.

Battilana, Julie, and Thomas D'Aunno. 2009. "Institutional Work and the Paradox of Embedded Agency." In *Institutional Work: Actors and Agency in Institutional Studies of Organizations*, edited by Thomas B. Lawrence, Roy Suddaby, and Bernard Leca, 32–58. Cambridge: Cambridge University Press.

Battilana, Julie, Bernard Leca, and Eva Boxenbaum. 2009. "How Actors Change Institutions: Towards a Theory of Institutional Entrepreneurship." *Academy of Management Annals* 3 (1): 65–107.

Bechky, Beth. 2011. "Making Organizational Theory Work: Institutions, Occupations, and Negotiated Orders." *Organization Science* 22 (5): 1157–67.

Berman, Elizabeth Popp. 2008. "Why Did Universities Start Patenting? Institution-Building and the Road to the Bayh-Dole Act." *Social Studies of Science* 38:835–71.

———. 2012. *Creating the Market University: How Academic Science Became an Economic Engine*. Princeton, NJ: Princeton University Press.

Binder, Amy. 2007. "For Love and Money: Organizations' Creative Responses to Multiple Environmental Logics." *Theory and Society* 36:547–71.

Boli, John, Francisco O. Ramirez, and John W. Meyer. 1985. "Explaining the Origins and Expansion of Mass Education." *Comparative Education Review* 29 (2): 145–70.

Bonikowski, Bart. 2016. "Nationalism in Settled Times." *Annual Review of Sociology* 42:427–49.

Boxenbaum, Eva. 2006. "Lost in Translation: The Making of Danish Diversity Management." *American Behavioral Scientist* 49 (7): 939–48.

Branscomb, Lewis M., Fumio Kodama, and Richard Florida. 1999. *Industrializing Knowledge: University-Industry Linkages in Japan and the United States*. Cambridge, MA: MIT Press.

Burgess, Chris. 2010. "The 'Illusion' of Homogeneous Japan and National Character: Discourse as a Tool to Transcend the 'Myth' vs. 'Reality.'" *Asia Pacific Journal* 8 (9): 1–24.

Cabinet Office of Japan. 2004. "The Minutes of the 18th Meeting of the Council for Science, Technology and Innovation (Draft)." Accessed May 18, 2022. https://www8.cao.go.jp/cstp/siryo/haihu19/siryo5.pdf.

Chorev, Nitsan. 2012. "Changing Global Norms through Reactive Diffusion: The Case of Intellectual Property Protection of Aids Drugs." *American Sociological Review* 77 (5): 831–53.

Cohen, Stanley N, Annie C. Y. Chang, Herbert W. Boyer, and Robert B. Helling. 1973. "Construction of Biologically Functional Bacterial Plasmids in Vitro." *Proceedings of National Academy of Science* 70 (11): 3240–44.

Colyvas, Jeannette A. 2007. "From Divergent Meanings to Common Practices: The Early Institutionalization of Technology Transfer in the Life Sciences at Stanford University." *Research Policy* 36 (4): 456–76.

Colyvas, Jeannette A., and Stefan Jonsson. 2011. "Ubiquity and Legitimacy: Disentangling Diffusion and Institutionalization." *Sociological Theory* 29 (1): 27–53.

Colyvas, Jeannette A., and Walter W. Powell. 2006. "Road to Institutionalization: The Remaking of Boundaries between Public and Private Science." *Research in Organizational Behavior* 27:305–53.

Cox, Rupert, ed. 2008. *The Culture of Copying in Japan: Critical and Historical Perspectives*. London: Routledge.

Cusumano, Michael A. 1988. "Manufacturing Innovation: Lessons from the Japanese Auto Industry." *MIT Sloan Management Review* 30:29–39.

Czarniawska, Barbara, and Guji Sevón, eds. 1996. *Translating Organizational Change*. Berlin: Walter de Gruyter.

D'Este, Pablo, and Markus Perkmann. 2011. "Why Do Academics Engage with Industry? The Entrepreneurial University and Individual Motivations." *Journal of Technology Transfer* 36 (3): 316–39.

DiMaggio, Paul J., and Walter W. Powell. 1983. "The Iron Cage Revisited: Institutional

Isomorphism and Collective Rationality in Organizational Fields." *American Sociological Review* 48 (2): 147–60.

Djelic, Marie-Laure. 2001. *Exporting the American Model: The Postwar Transformation of European Business.* Oxford: Oxford University Press.

Djelic, Marie-Laure, and Sigrid Quack, eds. 2003. *Globalization and Institutions: Redefining the Rules of the Economic Game.* Cheltenham, U.K.: Edward Elgar Publishing.

Djelic, Marie-Laure, and Kerstin Sahlin-Anderson. 2006. *Transnational Governance: Institutional Dynamics of Regulation.* Cambridge: Cambridge University Press.

Doi, Takeo. 1971. *Amae no Kouzou.* [Anatomy of dependance]. Tokyo: Koubundo.

Drori, Gili S., John W. Meyer, and Hokyu Hwang, eds. 2006. *Globalization and Organization: World Society and Organizational Change.* Oxford: Oxford University Press.

Drori, Gili S., John W. Meyer, Francisco O. Ramirez, and Evan Schofer. 2003. *Science in the Modern World Polity: Institutionalization and Globalization.* Stanford: Stanford University Press.

Emerson, Robert M., Rachel I. Fretz, and Linda L. Shaw. 1995. *Writing Ethnographic Fieldnotes.* Chicago: University of Chicago Press.

Emirbayer, Mustafa, and Ann Mische. 1998. "What Is Agency?" *American Journal of Sociology* 103 (4): 962–1023.

Etzkowitz, Henry. 1983. "Entrepreneurial Scientists and Entrepreneurial Universities in American Academic Science." *Minerva* 21:198–33.

———. 1989. "Entrepreneurial Science in the Academy: A Case of the Transformation of Norms." *Social Problems* 36 (1): 14–29.

———. 1998. *Capitalizing Knowledge: New Intersections of Industry and Academia.* Stony Brook, NY: SUNY Press.

———. 2002. *MIT and the Rise of Entrepreneurial Science.* London: Taylor and Francis.

Evans, James A. 2010. "Industry Induces Academic Science to Know Less about More." *American Journal of Sociology* 116 (2): 389–452.

Fligstein, Neil. 2001. "Social Skills and the Theory of Fields." *Sociological Theory* 19 (2): 105–25.

Fourcade, Marion. 2006. "The Construction of a Global Profession: The Transnationalization of Economics." *American Journal of Sociology* 112 (1): 145–94.

Friedland, Roger, and Robert R. Alford. 1991. "Bringing Society Back In: Symbols, Practices, and Institutional Contradictions." In *The New Institutionalism in Organizational Analysis*, edited by W. W. Powell and P. J. DiMaggio, 232–63. Chicago: University of Chicago Press.

Geuna, Aldo, and Federica Rossi. 2011. "Changes to University IPR Regulations in Europe and the Impact on Academic Patenting." *Research Policy* 40:1068–76.

Giddens, Anthony. 1979. *Central Problems in Social Theory: Action, Structure and Contradiction in Social Analysis.* Berkeley: University of California Press.

———. 1986. *The Constitution of Society: Outline of the Theory of Structuration.* Berkeley: University of California Press.

Glaser, Barney G. 1963. "The Local Cosmopolitan Scientist." *American Journal of Sociology* 63 (3): 249–59.

Glaser, Barney G., and Anselm L. Strauss. 1967. *The Discovery of Grounded Theory: Strategies for Qualitative Research*. New York: Aldine Transaction.

Gordon, Andrew. 2002. *A Modern History of Japan: From Tokugawa Times to the Present*. Oxford: Oxford University Press.

Gouldner, Alvin W. 1957. "Cosmopolitans and Locals: Toward an Analysis of Latent Social Roles I." *Administrative Science Quarterly* 2 (4): 444–80.

Greenwood, Royston, Christine Oliver, Roy Suddaby, and Kerstin Sahlin, eds. 2008. *The SAGE Handbook of Organizational Institutionalism*. Los Angeles: SAGE.

Greenwood, Royston, Mia Raynard, Farah Kodeih, Evelyn R. Micelotta, and Michael Lounsbury. 2011. "Institutional Complexity and Organizational Responses." *Academy of Management Annals* 5 (1): 317–71.

Greenwood, Royston, Roy Suddaby, and C. R. Hinings. 2002. "Theorizing Change: The Role of Professional Associations in the Transformation of Institutionalized Fields." *Academy of Management Journal* 45 (1): 58–80.

Haeussler, Carolin, and Jeannette A. Colyvas. 2011. "Breaking the Ivory Tower: Academic Entrepreneurship in the Life Sciences in UK and Germany." *Research Policy* 40 (1): 41–54.

Hafner-Burton, Emilie M., and Kiyoteru Tsutsui. 2005. "Human Rights in a Globalizing World: The Paradox of Empty Promises." *American Journal of Sociology* 110 (5): 1373–1411.

Hallett, Tim. 2010. "The Myth Incarnate: Recoupling Processes, Turmoil, and Inhabited Institutions in an Urban Elementary School." *American Sociological Review* 75 (1): 52–74.

Hallett, Tim, and Amelia Hawbaker. 2021. "The Case for an Inhabited Institutionalism in Organizational Research: Interaction, Coupling, and Change Reconsidered." *Theory and Society* 50 (1): 1–32.

Hallett, Tim, and Marc Ventresca. 2006a. "How Institutions Form: Loose Coupling as Mechanisms in Gouldner's Patterns of Industrial Bureaucracy." *American Behavioral Scientist* 49:908–24.

———. 2006b. "Inhabited Institutions: Social Interactions and Organizational Forms in Gouldner's Patterns of Industrial Bureaucracy." *Theory and Society* 35:213–36.

Haveman, Heather A., and Gillian Gualtieri. 2017. "Institutional Logics." In *Oxford Research Encyclopedias of Business and Management*. https://doi.org/10.1093/acrefore/9780190224851.013.137.

Healy, Kieran. 2005. *Last Best Gifts: Altruism and the Market for Human Blood and Organs*. Chicago: University of Chicago Press.

Henderson, Rebecca, Adam B. Jaffe, and Manuel Trajtenberg. 1998. "University Patenting Amid Changing Incentives for Commercialization." In *Creation and Transfer of Knowledge: Institutions and Incentives*, edited by Giorgio Barba Navaretti, Partha Dasgupta, Karl-Göran Mäler, and Domenico Siniscalco, 87–114. Berlin: Heidelberg.

Hirsch, Paul M., and Michael Lounsbury. 1997. "Ending the Family Quarrel: Toward a Reconciliation of 'Old' and 'New' Institutionalisms." *American Behavioral Scientist* 40 (4): 406–18.

Hughes, Sally S. 2001. "Making Dollars out of DNA: The First Major Patent in Biotechnology and the Commercialization of Molecular Biology, 1974–1980." *Isis* 92:541–75.

Inglehart, Ronald, and Pippa Norris. 2003. *Rising Tide: Gender Equality and Cultural Change Around the World.* Cambridge: Cambridge University Press.

Johnson, David R. 2017. *A Fractured Profession: Commercialism and Conflict in Academic Science.* Baltimore: Johns Hopkins University Press.

Kamatani, Chikayoshi. 2006. "University-Industry Cooperative Research in Japan: From Meiji Era to the Second World War" [Nihon ni okeru sangaku renkei: sono soushiki ni miru tokucho]. *Kokuritsu kyouiku seisaku kenkyujo kiyo* [National Education Policy Research Paper Series], 57–102.

Kameo, Nahoko. 2015. "Gifts, Donations, and Loose Coupling: Responses to Changes in Academic Entrepreneurship among Japanese Bioscientists." *Theory and Society* 44 (2): 177–98.

———. 2017. "A Culture of Uncertainty: Interaction and Organizational Memory in Software Engineering Teams under a Productivity Scheme." *Organization Studies* 38 (6): 733–52.

———. 2024. "Nationalizing Accounts: Everyday Nationalism, Japanese Scientists, and Global Policy." *Theory and Society* 53:167–92.

Kanama, Daisuke, and Kumi Okuwada. 2008. "A Study of University Patent Portfolios: Portfolio of Patent Application from Tohoku University." NISTEP (National Institute of Science and Technology Policy) Report. Accessed October 12, 2022. http://hdl.handle.net/11035/924.

Kelly, Erin, and Frank Dobbin. 1998. "How Affirmative Action Became Diversity Management: Employer Response to Anti-discrimination Law, 1961–1996." *American Behavioral Scientist* 41 (7): 960–84.

Kneller, Robert. 1999. "Intellectual property rights and university—industry technology transfer in Japan." *Science and Public Policy* 26 (2): 113–24.

———. 2003. "University-Industry Cooperation and Technology Transfer in Japan Compared with the Us: Another Reason for Japan's Economic Malaise?" *University of Pennsylvania Journal of International Economic Law* 24 (2): 329–449.

———. 2011. "Invention Management in Japanese Universities and its Implications for Innovation: Insights from the University of Tokyo." In *Academic Entrepreneurship in Asia: The Role and Impact of Universities in National Innovation System*, edited by Poh Kam Wong, 69–85. Cheltenham, U.K.: Edward Elgar Publishing.

Kneller, Robert, and Sachiko Shudo. 2008. "Large Companies' Preemption of University Inventions by Joint Research is Strangling Japanese Entrepreneurship and Contributing to the Degradation of University Science." *Journal of the Intellectual Property Association of Japan* 5 (2): 36–50.

Koyama, Kiyohito. 2017. "Japanese Culture and University-Firm Collaboration [Nihon-no bunka to sangaku renkei]." *Sangakukan renkei Janaru* [Industry-University-Government Collaboration Journal] 13 (11): 3.

Kreiner, Kristian, and Majken Schultz. 1993. "Informal Collaboration in R&D: The formation of Networks Across Organization." *Organization Studies* 14 (2): 189–209.

Lawrence, Thomas B., Roy Suddaby, and Bernard Leca. 2009. *Institutional Work: Actors and Agency in Institutional Studies of Organizations.* Cambridge: Cambridge University Press.

Ledford, Heidi. 2013. "Universities Struggle to Make Patents Pay." *Nature* 501 (7468): 471–72.

Leibel, Esther, Tim Hallett, and Beth A. Bechky. 2018. "Meaning at the Source: The Dynamics of Field Formation in Institutional Research." *Academy of Management Annals* 12 (1): 154–77.

Lim, Alwyn, and Kiyoteru Tsutsui. 2012. "Globalization and Commitment in Corporate Social Responsibility: Cross-National Analyses of Institutional and Political-Economy Effects." *American Sociological Review* 77 (1): 69–98.

Lounsbury, Michael. 2007. "A Tale of Two Cities: Competing Logics and Practice Variation in the Professionalizing of Mutual Funds." *Academy of Management Journal* 50 (2): 289–307.

Lukes, Steven. 1992. *Power: A Radical View.* New York: Palgrave Macmillan.

Mauss, Marcel. (1925) 1954. *The Gift: Forms and Functions of Exchange in Archaic Societies.* Translated by I. Cunnison. Glencoe: Free Press.

McPherson, Chad Michael, and Michael Sauder. 2013. "Logics in Action: Managing Institutional Complexity in a Drug Court." *Administrative Science Quarterly* 58 (2): 165–96.

Merton, Robert K. 1973. *The Sociology of Science: Theoretical and Empirical Investigations.* Chicago: University of Chicago Press.

Metlay, Grischa. 2006. "Reconsidering Renormalization: Stability and Change in 20th-Century Views on University Patents." *Social Studies of Science* 36 (4): 565–97.

MEXT (Ministry of Education, Culture, Sports, Science, and Technology, Japan). 2016. *Sangakukan renkei ni yoru kyoudou kenkyu kyouka no tame no gaidorain ni tsuite* [Regarding the guidelines for more effective industry-government-university research collaborations]. Committee on the Promotion of Innovation through industry-university-government collaboration. November 30. Accessed September 18, 2022. https://www.mext.go.jp/a_menu/kagaku/taiwa/1380912.htm.

———. 2017. *Daigaku tou ni okeru chiteki zaisan manajimento jirei ni manabu kyoudoukenkyuu tou seika no toriatsukai no arikata ni kansuru chosa kenkyu—introducing "Sakura tool"* [Introducing Sakura tool: Research on the management of the results of collaborative research and university IP management]. Accessed August 6, 2022. https://www.mext.go.jp/a_menu/shinkou/sangaku/1383777.htm.

———. 2003–2022. *Daigaku tou ni okeru sangaku renkei tou jisshi joukyou ni tsuite* [Yearly reports regarding the university-industry collaboration activities in universities]. Accessed October 18, 2022. https://www.mext.go.jp/a_menu/shinkou/sangaku/sangakub.htm.

Meyer, John W., and Brian Rowan. 1977. "Institutionalized Organizations—Formal Structure as Myth and Ceremony." *American Journal of Sociology* 83 (2): 340–63.

Meyer, John W., John Boli, George M. Thomas, and Francisco O. Ramirez. 1997. "World Society and the Nation-State." *American Journal of Sociology* 103:144–81.

Ministry of Education (Japan). 1962. *Nihon no seicho to kyouiku.* [Japan's development and education].

———. 1964. *Syougaku–kifu kin, Inin-keiri Jimu Toriatsukai Kisoku* [Regulations on the management of shogaku-kifu entrusted accounting]. Order no. 14, April 23.

———. 1978. *Kokuritu daigaku tou no kyoukan tou no hatsumei ni kakawaru tokkyo no toriatsukai ni tsuite* [Regarding the management of patents concerning the inventions of professors at national universities and equivalent institutions]. Notification no. 117.

———. 1984a. *Jyutaku Kenkyu oyobi minkantou tono kyoudou kenkyu no kakawaru tokkyo touno jissitou ni tsuite* [Regarding the use of intellectual property in relation to the sponsored or collaborative research with private corporations and similar entities]. Notification no. 172, May 8.

———. 1984b. *Shougaku kifu kin tou gaibu shikin no ukeire ni tsuite* [Regarding the acceptance of external funding such as research donations]. Notification from the director of international affairs and the director of the accounting office on December 22.

MITI (Ministry of International Trade and Industry, Japan). 2006. *Heisei 19 Nendo Sangyo-Gijyutu Tyosa: Gijyutu iten no tameno daigaku renkeigata IP fando keisei kanousei tyousa* [2017 Industry Technology Report: Regarding the possibility of funding for IPs emerging out of industry-university collaboration for technology transfer]. Accessed February 14, 2014. http://www.kanto.meti.go.jp/seisaku/gizyutsu/internship/data/19fy_ipfund/19fy_ipfund_hyoushi.pdf.

———. 2023. *Nihon-ban Bay-Dole Seido (Sangyo-Gijutsu Kyouka Hou clause 17)* [Japanese Bayh-Dole Act (Industry Technology Enhancement Act, clause 17)]. Accessed October 24, 2023. https://www.meti.go.jp/policy/economy/gijutsu_kakushin/innovation_policy/bayh_dole_act.html.

Mowery, David C., Richard Nelson, Bhaven Sampat, and Arvids A. Ziedonis. 2001. "The Growth of Patenting and Licensing by U.S. Universities: An Assessment of the Effects of the Bayh-Dole Act of 1980." *Research Policy* 30 (1): 99–119.

Mowery, David C., and Bhaven N. Sampat. 2001. "University Patents and Patent Policy Debates in the USA, 1925–1980." *Industrial and Corporate Change* 10 (3): 781–814.

———. 2002. "Learning to Patent: Institutional Experience, Learning, and the Characteristics of US University Patents after the Bayh-Dole Act, 1981–1992." *Management Science* 48 (1): 73–89.

Nelson, Andrew J. 2014. "From the Ivory Tower to the Startup Garage: Organizational Context and Commercialization Processes." *Research Policy* 43 (7): 1144–56.

Nelson, Richard R. 1993. *National Innovation Systems: A Comparative Analysis.* Oxford: Oxford University Press.

Odagiri, Hiroyuki. 1999. "University-Industry Collaboration in Japan: Facts and Interpretations." In *Industrializing Knowledge: University-Industry Linkages in Japan and the United States*, edited by Lewis M. Branscomb, Fumio Kodama and Richard Florida, 252–65. Cambridge, MA: MIT Press.

Office of the Cabinet, Japan. 2015. *Kuni no itaku ni yoru kenkyukaihatsu ni okeru chiteki zaisan manejimento ni kansuru unyoushishin no sakutei ni muketa torikumi jokyou ni tsuite* [Regarding the situation on the operation guidelines on the management of government-funded research-based IPs]. Accessed October 12, 2022. https://www.kantei.go.jp/jp/singi/titeki2/tyousakai/kensho_hyoka_kikaku/2015/dai4/siryou7.pdf.

Ono Yakuhin Kougyou Kabushiki Gaisha. 2018. *Koporate repoto 2018* [Ono Pharmaceutical Corporate Report 2018]. Accessed February 10, 2024. https://www.ono-pharma.com/sites/default/files/ja/ir/library/integrated_report/all_2018.pdf.

Owen-Smith, Jason. 2011. "The Institutionalization of Expertise in University Licensing." *Theory and Society* 40 (1): 63–94.

Owen-Smith, Jason, and Walter W. Powell. 2001. "To Patent or Not: Faculty Decisions and Institutional Success at Technology Transfer." *Journal of Technology Transfer* 26 (1–2): 99–114.

Parmentier, Richard J. 1994. *Signs in Society: Studies in Semiotic Anthropology*. Bloomington: Indiana University Press.

Pechter, Kenneth, and Sumio Kakinuma. 1999. "Coauthorship Linkages between University Research and Japanese Industry." In *Industrializing Knowledge: University-Industry Linkages in Japan and the United States*, edited by Lewis M. Branscomb, Fumio Kodama, and Richard Florida, 102–127. Cambridge, MA: MIT Press.

Perrow, Charles. 1986. *Complex Organizations: A Critical Essay*. New York: Newbery Award Records.

Powell, Walter W., and Jeannette A. Colyvas. 2008. "Microfoundations of Institutional Theory." In *The SAGE Handbook of Organizational Institutionalism*, edited by Royston Greenwood, Renate E. Meyer, Thomas B. Lawrence, and Christine Oliver, 276–98. London: SAGE.

Powell, Walter W., Kenneth W. Koput, and Laurel Smith-Doerr. 1996. "Interorganizational Collaboration and the Locus of Innovation: Networks of Learning in Biotechnology." *Administrative Science Quarterly* 41 (1): 116–45.

Rossman, Gabriel. 2014. "Obfuscatory Relational Work and Disreputable Exchange." *Sociological Theory* 32 (1): 43–63.

Sapir, Adi, and Nahoko Kameo. 2019. "Rethinking Loose Coupling of Rules and Entrepreneurial Practices among University Scientists: A Japan–Israel Comparison." *Journal of Technology Transfer* 44 (1): 49–72.

Scott, Richard W. 1995. *Institutions and Organizations: Ideas, Interests and Identities*. Thousand Oaks, CA: SAGE.

Scott, Richard W., and John W. Meyer. 1994. *Institutional Environments and Organizations: Structural Complexity and Individualism*. Thousand Oaks, CA: SAGE.

Selznick, Philip. 1949. *TVA and the Grass Roots: A Study in the Sociology of Formal Organization*. Berkeley: University of California Press.

———. 1996. "Institutionalism 'Old' and 'New.'" *Administrative Science Quarterly* 41 (2): 270–77.

Simmel, Georg. (1900) 2011. *The Philosophy of Money*. London: Routledge.

Slaughter, Sheila, and Larry L. Leslie. 1999. *Academic Capitalism: Politics, Policies, and the Entrepreneurial University.* Baltimore: Johns Hopkins University Press.

Slaughter, Sheila, and Gary Rhoades. 2004. *Academic Capitalism and the New Economy: Markets, State and Higher Education.* Baltimore: Johns Hopkins University Press.

So, Anthony D., Bhaven N. Sampat, Arti K. Rai, Robert Cook-Deegan, Jerome H Reichman, Robert Weissman, and Amy Kapczynski. 2008. "Is Bayh-Dole Good for Developing Countries? Lessons from the US Experience." *PLoS Biology* 6 (10): e262.

Stamatov, Peter. 2013. *The Origins of Global Humanitarianism: Religion, Empires, and Advocacy.* Cambridge: Cambridge University Press.

Stevens, Ashley J. 2004. "The Enactment of Bayh-Dole." *Journal of Technology Transfer* 29 (1): 93–99.

Stinchcombe, Arthur L. 1997. "On the Virtues of the Old Institutionalism." *Annual Review of Sociology* 23:1–18.

Strang, David, and John W. Meyer. 1993. "Institutional Conditions for Diffusion." *Theory and Society* 22 (4): 487–511.

Stuart, Toby E., and Waverly W. Ding. 2006. "When Do Scientists Become Entrepreneurs? The Social Structural Antecedents of Commercial Activity in the Academic Life Sciences." *American Journal of Sociology* 112 (1): 97–144.

Swidler, Ann. 2001. *Talk of Love: How Culture Matters.* Chicago: University of Chicago Press.

Tavory, Iddo, and Stefan Timmermans. 2014. *Abductive Analysis: Theorizing Qualitative Research.* Chicago: University of Chicago Press.

Thornton, Patricia H., and William Ocasio. 1999. "Institutional Logics and the Historical Contingency of Power in Organizations: Executive Succession in the Higher Education Publishing Industry, 1958–1990." *American Journal of Sociology* 105 (3): 801–43.

Thornton, Patricia H., William Ocasio, and Michael Lounsbury. 2012. *The Institutional Logics Perspective: A New Approach to Culture, Structure and Process.* Oxford: Oxford University Press.

Timmermans, Stefan, and Iddo Tavory. 2007. "Advancing Ethnographic Research through Grounded Theory Practice." Pp. 493–513 in *Handbook of Grounded Theory,* edited by Anthony Bryant and Kathy Charmaz. London: SAGE.

———. 2012. "Theory Construction in Qualitative Research: From Grounded Theory to Abductive Analysis." *Sociological Theory* 30 (3): 167–86.

———. 2020. "Racist Encounters: A Pragmatist Semiotic Analysis of Interaction." *Sociological Theory* 38 (4): 295–317.

Toole, Andrew A., and Dirk Czarnitzki. 2007. "Biomedical Academic Entrepreneurship through the SBIR Program." *Journal of Economic Behavior and Organization* 63:716–38.

Tsutsui, Kiyoteru. 2018. *Rights Make Might: Global Human Rights and Minority Social Movements in Japan.* Oxford: Oxford University Press.

Turco, Catherine. 2012. "Difficult Decoupling: Employee Resistance to the Commercialization of Personal Settings." *American Journal of Sociology* 118 (2): 380–419.

University of Tokyo, Division of University-Corporate Relations. 2011. *Minkan Kigyo-tono Kyoudou Kenkyu no arikata ni tsuite* [Report on the practices of collaborative research with private firms].

USPTO (United States Patent and Trademark Office). 2022. "U.S. Colleges and Universities—Utility Patent Grants, Calendar Years 1969–2012." Accessed December 12, 2022. https://www.uspto.gov/web/offices/ac/ido/oeip/taf/univ/org_fi/universities_f.htm.

Vaughan, Diane. 1986. *Uncoupling: Turning Points in Intimate Relationships*. New York: Oxford University Press.

Wang, Dan J. 2014. "Activating Brokerage: Skilled Returnees as Agents of Cross-Border Knowledge Transfer." *Administrative Science Quarterly* 60 (1): 133–76.

Weiss, Robert S. 1995. *Learning from Strangers: The Art and Method of Qualitative Interview Studies*. New York: Simon and Schuster.

Westney, D. Eleanor. 1987. *Imitation and Innovation: The Transfer of Western Organizational Patterns to Meiji Japan*. Cambridge, MA: Harvard University Press.

Willis, Paul. 1978. *Learning to Labour: How Working Class Kids Get Working Class Jobs*. New York: Columbia University Press.

Wimmer, Andreas, and Nina Glick Schiller. 2003. "Methodological Nationalism, the Social Sciences, and the Study of Migration: An Essay in Historical Epistemology." *International Migration Review* 37 (3): 576–610.

Yamaguchi, Yoshikazu. 2010. "Hojin-ka wo fukumu kikan ni okeru kokuritsudaigaku no gaibushikin ukeire no doukou no bunseki" [An analysis of the trends of the acceptance of external funds in national universities in the period including the transformation into independent corporations]. *Sangaku Renkei Gaku* 6 (2): 44–55.

Yoshihara, Mariko, and Katsuya Tamai. 1999. "Lack of Incentive and Persisting Constraints: Factors Hindering Technology Transfer at Japanese Universities." In *Industrializing Knowledge: University-Industry Linkages in Japan and the United States*, edited by Lewis M. Branscomb, Fumio Kodama, and Richard Florida, 348–64. Cambridge, MA: MIT Press.

Yoshino, Kosaku. 1992. *Cultural Nationalism in Contemporary Japan: A Sociological Enquiry*. London: Routledge.

Zelizer, Viviana A. 1994. *The Social Meaning of Money*. New York: Basic Books.

———. 2005. *The Purchase of Intimacy*. Princeton, NJ: Princeton University Press.

———. 2011. *Economic Lives: How Culture Shapes the Economy*. Princeton, NJ: Princeton University Press.

———. 2012. "How I Became a Relational Economic Sociologist and What Does That Mean?" *Politics & Society* 40:145–74.

———. 2013. *Economic Lives: How Culture Shapes the Economy*. Princeton, NJ.: Princeton University Press.

Zelizer, Viviana A. Rotman. (1979) 2017. *Morals and Markets: The Development of Life Insurance in the United States*. New York: Columbia University Press.

Zilber, Tammar B. 2002. "Institutionalization as an Interplay Between Actions, Meanings, and Actors: The Case of a Rape Crisis Center in Israel." *Academy of Management Journal* 45 (4): 234–54.

Zucker, Lynne G., and Michael R. Darby. 2001. "Capturing Technological Opportunity via Japan's Star Scientists: Evidence from Japanese Firms' Biotech Patents and Products." *Journal of Technology Transfer* 26 (1–2): 37–58.

Zucker, Lynne G., Michael R. Darby, and Jeff S. Armstrong. 2002. "Commercializing Knowledge: University Science, Knowledge Capture, and Firm Performance in Biotechnology." *Management Science* 48 (1): 138–53.

Zuckerman, Harriet. 1977. *Scientific Elite: Nobel Laureates in the United States*. New York: Transaction Publishers.

Index

CULTURE AND ECONOMIC LIFE

The Indebted Woman: Kinship, Sexuality, and Capitalism
Isabelle Guérin
2023

Traders and Tinkers: Bazaars in the Global Economy
Maitrayee Deka
2023

Identity Investments: Middle-Class Responses to Precarious Privilege in Neoliberal Chile
Joel Phillip Stillerman
2023

Making Sense: Markets from Stories in New Breast Cancer Therapeutics
Sophie Mützel
2022

Supercorporate: Distinction and Participation in Post-Hierarchy South Korea
Michael M. Prentice
2022

Black Culture, Inc.: How Ethnic Community Support Pays for Corporate America
Patricia A. Banks
2022

The Sympathetic Consumer: Moral Critique in Capitalist Culture
Tad Skotnicki
2021

Reimagining Money: Kenya in the Digital Finance Revolution
Sibel Kusimba
2021

Black Privilege: Modern Middle-Class Blacks with Credentials and Cash to Spend
Cassi Pittman Claytor
2020

Global Borderlands: Fantasy, Violence, and Empire in Subic Bay, Philippines
Victoria Reyes
2019

The Costs of Connection: How Data is Colonizing Human Life and Appropriating It for Capitalism
Nick Couldry and Ulises A. Mejias
2019

The Moral Power of Money: Morality and Economy in the Life of the Poor
Ariel Wilkis
2018

The Work of Art: Value in Creative Careers
Alison Gerber
2017

Behind the Laughs: Community and Inequality in Comedy
Michael P. Jeffries
2017

Freedom from Work: Embracing Financial Self-Help in the United States and Argentina
Daniel Fridman
2016

The authorized representative in the EU for product safety and compliance is:
Mare Nostrum Group
B.V Doelen 72
4831 GR Breda
The Netherlands

www.ingramcontent.com/pod-product-compliance
Lightning Source LLC
Chambersburg PA
CBHW030847270326
41928CB00007B/1254

* 9 7 8 1 5 0 3 6 4 0 4 0 5 *